# The Scientific Age

## THE IMPACT OF SCIENCE
## ON SOCIETY

*Based on the Trumbull Lectures
Delivered at Yale University*

# The Scientific Age

## THE IMPACT OF SCIENCE ON SOCIETY

### BY L. V. BERKNER

GREENWOOD PRESS, PUBLISHERS
WESTPORT, CONNECTICUT

Library of Congress Cataloging in Publication Data

Berkner, Lloyd Viel, 1905-
   The scientific age.

   Reprint of the ed. published by Yale University
Press, New Haven.
   Includes bibliographical references and index.
   1.  Science and civilization.  I.  Title.
[CB151.B39]1975]           501              75-16841
ISBN 0-8371-8263-8

Originally published in 1964 by Yale University Press, New Haven

Reprinted with the permission of Yale University Press

Reprinted in 1975 by Greenwood Press,
a division of Williamhouse-Regency Inc.

Library of Congress Catalog Card Number 75-16841

ISBN 0-8371-8263-8

Printed in the United States of America

*To Lillian Frances Berkner*
*my companion in the far corners of my laboratory—*
*Planet Earth*

# Contents

# List of Figures

# List of Tables

# Preface

In undertaking the Trumbull Lectures at Yale University for 1964, I chose to discuss the interaction of science and society—an interaction now strongly manifested through the technological revolution that went into high gear at mid-century.

Had I limited myself to my present field of scientific inquiry—the origin and history of the earth's atmosphere—the discussions would probably have been far more entertaining, and certainly easier for me. To range beyond one's field of special scientific competence has obvious dangers. Yet the magnitude of the social phenomena arising out of today's virile science and technology is such as to require study and analysis from every quarter. Too often in history great opportunity has been lost when men failed to grasp or even to recognize that opportunity, or to appreciate the crisis of which they were a part. In a rapidly moving social drama, the probability of misinterpreting its real meaning while we cling to outmoded doctrine or tradition is very great.

I was prompted to choose this subject for the Trumbull Lectures, in part, because of my recent work in Texas. The true import of the technological revolution is not very apparent to those in the traditional industrial communities where its progress has been more gradual and more orderly. But to one who came to the Southwest after having lived in the northeastern part of the United States for forty years, the immensity of the social change occurring along the great frontiers of our nation is almost staggering. Here, within a single generation, a whole population—indeed half our entire American population—is being forced from a basic dependence on agriculture and natural resources to a quite new dependence on manufacture and industry. Consequently, the leaders at these frontiers are acutely sensitive to the magnitude of the adjustments required and to the character of the new economy dictated by the pre-existing competition from the older, stable industrial areas of the nation.

Since graduate education plays a distinctive, even a key, role in this revolution, I am prompted to recall that it was at Yale in 1861 that the first three doctorates in philosophy in the United States were conferred. In the intervening hundred or so years, this precedent at Yale has established a trend in education that exercises a decisive influence on our present national situation.

One of those first three Ph.D.s in 1861 was earned by Arthur William Wright whose field was science—more specifically, physics and astronomy. He was a scientist of broad interests in mathematics, mineralogy, botany, and modern languages and was one of the revisers of Webster's *Dictionary;* he was also admitted to the bar. His early investigations of X rays gave the first definitive results in America with this powerful new physical tool. As a professor at Yale he built and occupied the first Sloane Laboratory of Physics. The catholicity of Wright's interests has been an example followed by leading scientists down to this day.

One recalls, also, the great J. Willard Gibbs, whose historic invention of quaternions gave to mathematical physics the powerful tool to manage vectors and tensors—the matrix theory that was the foundation for the work of Mach, of Minkowski, of Lorentz, and of Albert Einstein—indeed the basis for much of modern physics. I believe Gibbs was the second Yale Ph.D. in physics and the fifth of Yale University; he was certainly among the most illustrious of America's men of science.

Since it is obviously impossible to touch upon all of the ramifications of the powerful interactions of science with society, I shall consider first the significance of science and technology as the base of our national economy as a whole. Then I propose to discuss the principal points of impact on that new economy which determine its ultimate stability and effectiveness: (1) education in a new age; (2) government in relation both to education and to scientific research; and (3) the traditional aspects of the humanities and liberal arts in relation to the emergent society. From these special situations, I hope there will emerge a fairly clear pattern of what we must do if we are to realize for ourselves, and indeed for mankind, the full benefits of this remarkable efflorescence of intellectual activity.

It will be apparent as I go along that I have limited my observations rather severely to the impact of science and technology upon the United States, which more than any other nation has harnessed the new forces to achieve heretofore undreamed heights of productivity and wealth. Elsewhere in the world, of course, the situation is vastly different.

One can find no more graphic demonstration of the effects of the technological revolution upon a society capable of receiving it than to contrast the economic reviews of the several continents, so effectively presented each year by the

*New York Times.* This year, for example, the *Times* summarized our economy[1] as follows:

> The United States appears poised on the brink of a great new economic era. . . . Fueled by the expanding needs of a growing population, the nation's economic engine in 1964 is expected to churn out, for the first time, a volume of goods and services exceeding $600 billion . . . a rather amazing growth rate in the vicinity of 6 per cent. . . . Profits last year topped all records in American corporate history.

In Latin America, by contrast, the *Times*[2] noted an average per capita annual income of $325 in 1962 and observed that "these stagnating trends were still in evidence in 1963."

In the review of Asia[3] we find the depressing intelligence that "the peoples of Asia lost ground last year in their fight against hunger and poverty. Despite this fact, many leaders permitted politics and chauvinism to aggravate the economic problems of the area." The situation in Africa[4] was characterized as having "the quality of a telecast from an erratic time-machine" with important "firsts" in some countries taking the form of the erection of a shoe factory or bicycle plant, the building or paving of roads, or the beginnings of programs to increase the use of fertilizers[5] and more modern farming techniques.

1. Jan. 6, 1964, p. C49.
2. Jan. 17, 1964, pp. C45, 61.
3. Jan. 13, 1964, p. 37.
4. Jan. 20, 1964, p. 45.
5. This simple and disarming statement illustrates a typical failure to take into account the scientific problems that underlie man's improvement. For example, fertilizer is only effective when altogether new seed strains are developed to match the properties of the suitably fertilized soil, to say nothing of climate, as the Russians are learning to their disgruntlement.

Obviously these wide variations in the human condition are of vital concern to us from both a humanitarian and an economic and political point of view. But if we are to prepare ourselves to deal intelligently with the relatively hopeless situation in half the world, we can perhaps wisely begin by endeavoring to analyze more clearly the dominant elements of our own economic and social revolution in this technological age. As we see these in more adequate perspective, we may begin to perceive with fuller understanding the relative impact of the underdeveloped world on the stability and potential progress of our society and ultimately, perhaps, the means whereby the underprivileged world can achieve its own opportunities.

In preparing these lectures, I am indebted to my many colleagues in the Southwest who have aided, through their penetrating discussions, in clarifying and sharpening the issues that face us today. It would be impossible to mention all whose thinking has stimulated my own. Yet in particular, the discussions of my colleagues on the Executive Committee of the Graduate Research Center of the Southwest— Mr. Erik Jonsson, Mr. Eugene McDermott, Mr. Cecil Green, Mr. Stanley Marcus, Mr. Earl Cullum, Mr. C. A. Tatum, and Mr. Ralph Stohl—have provided a discipline that forces one to penetrate to the root of each issue.

I am indebted to Mrs. Dee Bartley and to Miss Susan DeWees for their untiring assistance in preparation of the manuscript and for their assistance in the search for reference material. In particular, I express my gratitude to Miss Lee Anna Embrey for her critical and analytical review of the manuscript. Through her long association with scientific leaders in government, beginning with Vannevar Bush in the days of the OSRD, extending through the organization with me of the Research and Development Board of the Department of Defense in 1947 and of the Science Program of the Department of State in 1950, and during her long tenure at the National Science Foundation, Miss Embrey has developed deep insights that have encouraged me to clarify my thoughts and phraseology. Her suggestions have been invaluable.

LLOYD V. BERKNER

*Fort Lauderdale*
*February 1, 1964*

Oh Science, lift aloud thy voice that stills
The pulse of fear, and through the conscience thrills—
   Thrills through the conscience the news of peace—
How beautiful thy feet are on the hills!

*Titus Lucretius Carus* (as translated
by W. H. Mallock)

# 1. The Economy of Plenty

My colleague of the Antarctic exploration, the distinguished geophysicist Jock Tuzo Wilson, once remarked:

> Up to the present, scientists generally have felt that their proper concern was with science, with the discovery of facts about nature. Their place was in the laboratory or, if they were in government or industry, as specialists to be called upon when needed, but not to be incorporated into executive ranks.
>
> When there were only a few scientists, who knew relatively little about nature, and had little influence on society, this attitude was probably sound enough. But times have changed. There are now a great many scientists. They have profoundly affected our civilization. If the ideas and tools of modern science are to be fully used to benefit mankind, scientists will have to consider social problems.[1]

With this encouragement, I propose to examine certain interactions of science with society in our day. One approaches a subject of such magnitude with a sense of brashness, because of its utter complexity. Yet the powerful de-

1. J. Tuzo Wilson, "The Scientist and Society," *Imperial Oil Review, 47,* No. 6, Dec. 1963.

velopments of our scientific age must be widely understood if civilization is to continue with a "positive derivative." So we shall begin by trying to isolate, identify, and characterize some of the "first order" effects on society produced by the revolution in science and technology.

Let us seek to identify the principal features underlying our economic development. What are the major events of our time that are new to the world? First, our populations are exploding at a rate approximating 2 per cent per year. Whereas the doubling time for our population was 200 years at the time of the American Revolution, in this century it is only 40 years. Science joined to medicine has extended the average life expectancy beyond the Biblical three score years and ten.

Second, at midcentury, agriculture is almost completely mechanized and industrialized. With the application of biology, chemistry, and mechanics to farming, agricultural productivity has increased a hundred times. A tiny fraction of our population is able to produce enough food to meet our own requirements, with great surpluses and the resulting problem of what to do with them. Whereas 70 per cent of the American population was engaged in agricultural pursuits in 1860, a century later only 7 per cent was so employed.

Third, the same situation obtains in the natural resources industries—fuels and minerals. The new technology has enormously enhanced the production of these basic commodities.

Fourth, a growing trend toward urbanization, linked to an exploding population, tends to concentrate the bulk of our people in just 100 or so major metropolitan areas which are bursting at the seams. In Texas our population is increasing 25 per cent per decade. Yet of our 254 counties, all but 31 have either lost population or remained about the same in each of the last three decades. Since 1950 more than half a million Texans have left the farming areas. Only nine or ten great metropolitan areas of Texas are absorbing the flow of migrants from the land, who, in addition to the newborn, are increasing their populations more than 50 per cent per decade. The example in Texas is typical of the whole nation.

Fifth, the productivity of the industries that support the population of these cities is gaining enormously. A powerful technology, sustained by an equally powerful and progressive science, makes possible the automatic control of factories and even whole industries. Fewer, but very skilled, men and women are required to produce larger quantities of superior products and services.

Sixth, the standard of living of the educated, skilled, and productive elements of our population is constantly rising. On October 7, 1963, the *New York Times* noted:

> The family budget standard just issued by the Community Council shows that New Yorkers in the low-to-moderate income brackets are eating more meat, seeing more shows, walking on thicker rugs, and generally living better than they did 8 years ago.

Even the poverty against which President Johnson has recently declared "all-out war" is by no means widespread but is concentrated in so-called "pockets of poverty."

Seventh, chronic unemployment is nevertheless increasing. The same issue of the *Times* that reported on new levels of prosperity in this country also noted on the same page:

> The central fact of life on the labor front was the continuing high level of unemployment. The year ended as it began, with the unemployment rate pushing 6 per cent of the labor force.

This is the inevitable consequence of the sudden rush of population to the cities and the coincident rise in our capac-

3

ity to produce more goods with fewer people as a result of constantly improving technology.

These then are the basic elements of our social and economic situation. We are witnessing a technological revolution derived from the powerful science of our time, leading to vastly improved productivity and greater wealth, but accompanied by a mass movement of people away from the land and concentrating in a few American metropolitan areas, and a steadily rising rate of unemployment, primarily among the uneducated.

We have come a long way from the traditional "economy of scarcity," which has plagued mankind since the beginning of history, into an "economy of abundance." For the first time, man has acquired, out of science and technology, the power to produce adequately to supply all his needs and a significant surplus as well. We must recognize, however, that the limitation of this capacity to the highly industrialized nations of the West is responsible for many of the world tensions with which political leaders grapple.

Since the beginning, man has struggled for the bare necessities—for the power to open his planet for his development and benefit. Then suddenly, at midcentury, he finds that he has ready access to the whole planet and, in fact, is able to escape from it completely.

Up to a hundred years ago, man's physical progress was characterized primarily by small improvements in his methods of obtaining the basic necessities—food, clothing, and shelter. In the present century, as we move toward the new science-oriented economy of plenty, our progress is characterized by radically new and powerful *industries* that are *not directly related to basic necessities*.

Electric power, the automobile, aircraft and other fast transportation, communications and mass media, the computer and automatic control, the nuclear and the space industries are typical examples. These new "adaptive" industries have the characteristic not of providing necessities but of making the individual far more effective in his environment. And these adaptive industries are characteristically created out of innovation derived from the most advanced science of our time.

The ultimate growth of the "industries of necessity" is determined by population. There is a limit, within a factor of two or three, to the amount of food, clothing, and shelter required. But, speaking generally, there is no foreseeable limit on the ultimate growth of the adaptive industries— those that better fit man for the command and enjoyment of his environment. The growth rate of adaptive industry is contingent only on the availability of scientific innovation, on the creation of technologies that can employ innovation in suitable and effective form, and, perhaps most important of all, on the motivation of people to accept as necessity those innovations that enlarge their horizons and increase their control over their environment.

It is becoming quite clear that in our own society the industries related to basic necessities no longer have the growth potential needed to overcome our increasing unemployment. We must look to the adaptive industries to provide new products and services and, with them, new sources for employment, wealth, and human satisfaction.

Here we come to the crux of the situation. New industry can be derived only from innovation. Innovation—what the late Sumner Slichter[2] called "the industry of discovery"— is the altogether new resource that already sustains more than half our economy. Innovation in turn depends predominantly on an ever-expanding science, guided by men educated and trained to enlarge that science significantly

2. National Science Foundation, *Proceedings of a Conference on Research and Development and its Impact on the Economy* (NSF–58–36), Washington, 1958, p. 109.

5

and to transform it into new and useful products and services.

In speaking of technology derived from science, we must recognize that technology has sprung from two quite different origins. When there is a job that must be done, the technologist must proceed with whatever knowledge, experience, materials, and tools he finds at hand. The building, road, or bridge to be built, or the needs of the deceased or dying, cannot await the development of more useful concepts. So technology had its origins in the gross and uncritical synthesis of experience, with little scientific theory to guide it. Thus early technology was represented in the mechanics arts as practiced by guilds or highly specialized groups of men who passed on their accumulated skills and judgments largely to apprentices.

But with the introduction of science, which generalizes experience in theoretical and more broadly conceptual form, a new foundation for technology emerged—withal slowly at first. From the theories of Galileo and Newton a more formalized mechanics appeared, to guide the more efficient and effective engineering of structures. The advances in mathematics of Descartes and Leibniz and the discovery of vector analysis provided powerful new tools for technologic designs. The formulation of conceptual chemistry by Priestley, Lavoisier, and their successors and the development of molecular concepts in subsequent years provided the scientific base for the technology of materials, metallurgy, and the chemistry of complex compounds. The interaction of the technologic concept of Watt's steam engine with the theoretical thinking of Helmholtz, Mayer, and Carnot led to the conceptions of thermodynamics which made possible the efficient power of the railway and the steamship and paved the way for the development of the internal combustion engine, the turbine, and now the rocket engine. The experimental researches of Faraday and Henry

6

had to be completed before electric power, the telegraph, and the telephone could be conceived. Radio could not be invented until the highly theoretical concepts of Maxwell provided for Hertz the experimental base upon which to demonstrate controlled electromagnetic waves. To cap this very incomplete sequence, the wholly theoretical constructs of Einstein's special relativity were required to relate mass and energy before man could conceive of nuclear energy.

The present century has been especially fruitful in the advances of science that encourage technological advance. Planck at the turn of the century conceived energy as flowing in discrete packages, a theory which received confirmation from Einstein's interpretation of the photoelectric effect in 1905. Rutherford isolated the atomic nucleus in 1911, and Bohr described the atom in 1913 as the nucleus enclosed in electron shells. Immediately the basic phenomena of chemistry, spectra, and radioactivity came together in these broader conceptual constructs. Einstein announced general relativity in 1920, and in 1924 Heisenberg stated the principle of uncertainty beyond which cause and effect could have no meaning.

In 1926 de Broglie formulated the wave theory of matter, and immediately Schroedinger and Dirac created the wave mechanics in statistical form. The neutron was discovered by Chadwick in 1932, thus completing the catalogue of stable nuclear particles, and in 1938 Hahn and Strassman demonstrated nuclear fission. Our century has given us the useful science of the microcosm, identifying the basic building blocks of nature—theoretical constructs that permit us to deal intelligently with matter.

To these achievements we should add the formulation of information theory by Shannon in 1948. From this wide variety of scientific discoveries in our present century, technology has acquired tremendous power. Electronics has made possible new forms of communication and navigation;

and semiconductors, evolved out of pure theory, have been applied to the development of miniature but sophisticated computers—devices that permit man to exercise unlimited control at great distances. An immense variety of materials is emerging out of extreme high and low temperature techniques, magnetic phenomena, and advanced chemical researches. We are nearing a threshold where control of materials may give us almost any desired physical property on command.

Molecular biology and genetics have provided, besides a new comprehension of biological structure and disease, almost certainly the ultimate chemical control in manufacture of products, not only in the perfection of their material content but also *in their form,* just as nature now accomplishes the task of replication through genetic coding. Moreover, these developments have opened innumerable new avenues for scientific investigation, many of which will offer still further opportunities for technology. Thus, step by step, the foundations of technology are being transformed from their original and wholly uncritical empirical base to the conceptual and highly theoretical base of pure science. Without this new foundation the technologies of tomorrow could not even be dreamed.

Within the next generation, or by the end of the century, some 35 years hence, technology will have become almost completely science-based, deriving its strength from the theoretical constructs and basic generalizations of science. The crossover from empirical to science-based technologies occurred about the midcentury, and now they constitute more than half the whole.

Already our schools of engineering are changing their character to accept the challenge of providing access to science in great depth. The innovations of the future will be derived from the broad precepts and generalizations of science as they emerge. In this perspective we see scientific re-

8

search, supported and conducted on an adequate scale and in a creative atmosphere, together with a science-oriented technology, as the wellsprings from which major new industry will arise. New enterprises are expanding at twice the growth rate of traditional industry; already they have become the principal source of new employment for the men and women crowding into our metropolitan areas. Of course our total production of goods and services cannot continue indefinitely to rise exponentially, but the growth that is provided by our new resource can bridge the gap between an economy of scarcity and an economy of plenty. In the transition, the rate of productivity can be multiplied many times to satisfy the world's needs.

The last century saw a vast reduction in American poverty as a result of the improved productivity acquired from our new adaptive industries and their exploitation of scientific opportunity. Our remaining "pockets of poverty" await the expansion of this process to a point where production surpluses can compensate for those small unproductive elements of our society that will probably always be with us.

As a nation, we are now fully committed to this new resource for our future national growth. As the productivity of highly educated and creative individuals increases, however, so does the number of people in the labor market. Therefore, full employment, with the consequent maximization of total wealth and elimination of poverty, requires the continuous introduction of new products and services that people need and will buy. Thus there must also be a steady source of "innovation" upon which new industries can be based.

Automation improves productivity, lowers prices, and gives us more useful "quality" products but throws men out of work. Then we look to science for the innovation with which to create new industries with new opportunities

9

to employ the unemployed. Thus science and technology have plunged us into a never-ending race between automation and innovation, a race to keep our economy healthy. The reward for pushing innovation ahead is ever-increasing and more broadly based wealth, individual opportunity, and community satisfactions. Once embarked in the race, there is no turning back. Consequently, we are forced to examine the character of the resource in detail so that we understand it and can develop it in optimum form.

As I see it, the successful manipulation of this new resource and the ultimate stability of the economy based upon it are dependent on five very basic elements:

1. Scientific research, to discover the functional behavior of nature in all its aspects, the general laws that govern that behavior, and the technologies that will establish man's beneficial control of nature through a growing range of methods.

2. Education, very advanced and continuing—at the graduate level and beyond—to further knowledge in all areas through scientific research and to see in the new knowledge innovations potentially useful to mankind. Sound education below the graduate levels is also imperative, both to provide the advanced skills needed in highly technological industries and to prepare the most competent students for still higher education.

3. Energy, derived from and controlled by the technologies as they emerge. Today, controlled energy replaces "labor" in the traditional sense.

4. Capital, which diverts a suitable proportion of the energies of society toward a highly sophisticated mechanization of the means of production, so that productivity is greatly enhanced and the power of the new resource thereby fully realized.

5. A motivated buying public that sees in entirely new

products or services the possibility of improved adaptation to the environment. Innovation is sterile unless it creates products in a form and under circumstances that will "sell."

In reviewing this list—scientific research, education, controlled energy, capital, and a public motivated to "need"—one is struck by the thought that the economy of plenty is governed by quite new and different social and economic doctrines. For example, we all learned as youngsters from Adam Smith that "labor is of equal value to the laborer" and that "labor is the basis for all value." In this economy of plenty where controlled energy and highly developed applied brainpower are the principal bases of value, labor contributes only in proportion to its manipulative power over nature. Likewise the words *buy* and *sell* acquire a somewhat different connotation as they relate to the health of the whole society.

Now any such assertions that run head-on into centuries of theory, statements that at most are applicable only to the United States and perhaps to other limited regions of the West, are likely to be received with a certain skepticism and even derision. The humanitarians will observe that they have little relevance to the economic and social conditions of more than half the world. But such criticisms miss the main point. We *have* created since the midcentury, in at least one nation, an economy that comes close to the elimination of poverty. This economy is based on no simple theory but on the basic premise that in any complex human situation, beyond the reach of traditional measures, one must resort to a system of diversity from which natural selection tests and selects the most suitable solutions. This system of diversity, or free enterprise if you like, is protected by a constitutional philosophy. From this basic premise has emerged the most promising economic system devised to date, a system of many defects, but nevertheless the only system that has brought men as a group to high economic achievement—an achievement that is at the same time ultimately identified with man's intellectual goals.

The outlines of this new economy cannot be forced into the framework of now outmoded economic and social theory. Growing as it does out of the recent development of science and technology, its principles have not yet been fully apprehended. After living in an economy of scarcity for five millennia, it is going to take some time and some creative imagination to adjust our thinking to this new economic base where brainpower, controlled energy, and a public determination of need are the major contributors to value.

One is led to more than a suspicion that this new economy of plenty is governed by doctrines quite different from those in the familiar but now obsolete economy of scarcity. Before indulging in theoretical speculations about some ideal economy of the future, we might better bluntly ask ourselves "How does our successful new economy of plenty really work?" and then generalize from there. For example, education, once a convenient stepping-stone to success, now becomes a central contributor to value. In particular, graduate education to the doctor's degree and beyond, only recently an intellectual luxury or the indispensable "union card" for teachers, is now vitally necessary to the success of the new economy and the society depending on it. The key point here is that a suitable mix of men, educated to command the limits of knowledge, is essential to the generation of adequate employment opportunity for all the others. This is an altogether new phenomenon in society.

To illustrate: in 1950 we employed a total of about 100 Ph.D.s in the Dallas–Fort Worth metropolitan area, most as teachers and professors. Today, slightly more than a decade later, this number has grown to 1,200. It is increasing by 200 per year. The population of this one great metropolitan area has grown to nearly 2,000,000, with Dallas County alone increasing by nearly 60,000 per year. The new industry in the area is almost wholly science-based, requiring about one science-trained Ph.D. to provide opportunity for each 115 others employed. Since all Texas universities combined cannot now meet the needs for trained scientists in even this one of its ten metropolitan areas, we can see the immense problem in advanced education that lies ahead.

The science and engineering education that met the purposes of my own generation can no longer command the ideas that will underlie engineering innovation of tomorrow. The days of the individual innovators—the Edisons and the Fords—have been superseded by the power of wave mechanics, stochastic processes, and advanced circuit theory. Whitehead observed that

> Inventive genius requires pleasurable mental activity as a condition for its vigorous exercise. "Necessity is the mother of invention" is a silly proverb. "Necessity is the mother of futile dodges" is much nearer to the truth. The basis of the growth of modern invention is science, and science is almost wholly the outgrowth of pleasurable intellectual curiosity.[3]

To capture and control the potential of the new technology, to create industry from it, and to direct it for our national benefit, we must have men who understand and exploit the potentialities of the fundamental science from which it is generated. One can cite repeated examples of fine factories, equipped with the most modern devices, well financed, directed by sincere and industrious men, whose wheels have stopped turning for want of one, or two, or perhaps a half dozen scientific and engineering leaders who

3. A. N. Whitehead, *The Aims of Education,* Macmillan, New York, 1929, p. 69.

could solve a critical technological problem. In the competitive marketplace, success, profits, and price reduction are increasingly dependent on the quality of the company's assets of intellectual strength in science and technology.

Indeed, qualified technologists are more and more appearing as corporate heads, or as members of corporation boards. I personally know of one very large company in which a major capital appropriation by its directors rested on their assessment of an abstruse problem of wave mechanics. To conceive new and improved products, to make decisions promptly, to finance them suitably, and to avoid costly mistakes in the competitive new world requires competent, science-oriented technologists at every level of industrial operations. The radically changed demands upon the labor force for skills of an extraordinary kind raise a critical problem in meeting them, and this problem will be considered in some detail in the next chapter.

Still other issues offer fruitful opportunity for study and speculation. Labor leaders already differ sharply on the benefits of automation. Meany considers automation immoral, while Reuther hails its potential ability to eliminate poverty. Reuther is probably right, but we have yet to solve the problem of how to employ usefully the added time that automation makes available. To the suggestion of merely shortening the work week without collateral social adjustments, the New York taxi driver will reply out of firsthand experience, "Yeah, and let Mike Quill's busters drive taxis the other three days to kill our own jobs!"

Quite clearly, man's new leisure will at the same time require some form of constructive outlet. Since education can never be said to be complete, various forms of continuing and formal education doubtless will develop as highly useful social concepts. A more profound need, however, is for the emergence of new moral concepts to govern relationships among men in this quite new environment in which the proportion of time devoted to labor and to leisure has been reversed and in which the concept of leisure is modified. For in our new society each individual will be constrained to rise to the level of his full capability and to employ a suitable proportion of his new leisure to that end.

So in codifying the doctrines governing our new economy, we must examine the real issues, such as the proportion of the population that can be highly trained and educated, the most suitable concept of leisure required to make that economy optimally viable, and how that leisure ought to be employed.

Even more critical is the need to identify the elements of instability in the new economy—instabilities introduced by serious inadequacies in education at particular levels, by unsuitable levels of support of the science from which innovation is derived, by the growing inequality of the "have" and the "have-not" nations, and above all by failure to identify the prime sources of stability, as well as instability, so that society can play its role intelligently. I shall touch on some aspects of these subjects in subsequent chapters.

Since science is at the root of this situation, it may be useful to summarize the points of impact at which scientific research is radically modifying our social situation. I would emphasize five:

First, the growing technological capability derived from an ever-increasing knowledge of nature. As he solves the secrets of nature, man acquires greater control, thereby adding to his effectiveness in his environment. The industrial opportunity that science thus provides has scarcely been tapped, and we are on the threshold of immense future opportunities if we but maintain a vital and viable scientific effort. We are certainly within reach of abolishing poverty, with its degradation of the human spirit.

Second, the central role that scientific research must play in the graduate and the postdoctoral training of scientists.

15

We cannot maintain an adequate level of sound scientific research without a strong body of highly trained and creative scientists. You do not become a highly skilled pilot by reading about flying in books; you first fly in airplanes with competent instructors and then perfect your own skills by endless hours of solo flight. In the same way, you don't become a scientist merely by attending classes or reading books. You work in the laboratory with experienced professors on creative and original problems to learn how to deal with novel situations and to discover and recognize the significance of new knowledge. Parenthetically, I would add that just as the training of pilots involves expensive duplication, so we must recognize that training of scientists likewise justifies much duplication in research effort. Proficiency requires not only the underlying book knowledge but also the dexterity acquired from actual participation in all phases of the scientific process.

Third, scientific research influences public attitudes toward new ideas, products, and services. No matter how useful a new idea or device may be, it will not create employment, or wealth, or opportunity, unless the public will accept and buy it. The spotlight on scientific research, and the wonders of nature it discloses, unquestionably influences the public to acceptance of the new products and services that a science-oriented economy can provide. Our space program has an enormous impact in this respect. It lifts people from the horse-and-buggy age and accustoms them to the ideas of change and progress. It motivates them to a more sophisticated and effective enjoyment of their environment.

Fourth, the advancement of knowledge always influences civilization in the cultural sense. The incredible brutality of the past, so aptly described by Hans Zinsser in *Rats, Lice, and History,*[4] is no longer tolerated in our own society,

4. Little, Brown, Boston, 1935.

when knowledge coupled with justice has lifted our civilization above the levels of animal-like existence. Knowledge enriches life and increases our ability to enjoy and appreciate the opportunities that our culture affords.

Finally, we cannot ignore the powerful relation of scientific research to the defense of our nation. Only when we are in command of the hinterlands of thought with its advanced weaponry can we guard against an enemy immobilizing us with some surprise discovery.

Before concluding this chapter, I should like to refer briefly to the third of my five points, namely, the impact of science on public attitudes toward the new ideas, products, and services so essential to a healthy economy during the next century. The phenomenon of the *Zeitgeist* is not susceptible to scientific analysis, but it is very real.

In his discussion of the Century of Genius—the sixteenth —Sir James Jeans observed:

> It would be very undiscerning to suppose that such a period of greatness could arrive as a mere accident, a specially brilliant galaxy of exceptional minds just happening to be born at one particular epoch. . . .
>
> The laws of probability will see to it that no abrupt jump [of mental ability] occurs from one generation to the next. Thus a period of greatness must be attributed to environment rather than accident; if an age shows one particular form of greatness, external conditions must have encouraged that form. For instance, the Sixteenth Century was an age of great explorers because conditions then favored exploration; the pioneering voyages of Columbus, Vasco da Gama, Cabot, Magellan and others had drawn attention to the wealth of new territory awaiting discovery, while men had learned to build ships that could defy the worst fury of the ocean.[5]

5. Sir James Jeans, *The Growth of Physical Science,* Cambridge University Press, 1951.

The emergence of the spirit of science, indeed the founding of the new sciences in the sixteenth century, is described by Wolf as "a big stride towards a freer and fuller rationality, unrestrained by arbitrary barriers."[6]

In our own day we must make decisions regarding major programs involving vast expenditures for the advancement of science and technology. Typically, consider the problem of sending men to the moon. What are the really compelling factors on which a proper judgment should be rendered? Among our national leaders, both in and out of science, one can find all shades of opinion, based on widely differing points of view.

On the side of the pros, represented by Harold Urey, Harry Hess, Frederick Seitz (President of the National Academy of Sciences), and myself among others, the arguments can roughly be summarized:

1. The opportunity to visit the moon and the planets is the most exciting scientific prospect of our time or perhaps any time.

2. Space is a symbol of technological superiority, and our loss of it might encourage the Russians to irresponsible adventure.

3. To be safe militarily, we must fully understand the capabilities of space and the elements of its control.

4. The present moon program with its well-defined goals focuses attention on the essentials. To delay will cost more, perhaps much more, in the long run.

5. Space technology is technology at its present outer limits, and will produce vast increments in a variety of technological gains and innovations.

6. New symbols of human progress, such as the conquest

6. A. Wolf, *A History of Science, Technology, and Philosophy in the 16th and 17th Centuries,* The Science Library, Harper, New York, 1950 (reprinted 1959).

of space, stimulate people to accept more advanced and imaginative forms of living, including new and advanced types of consumer goods.

7. Any national goal that will increase man's effort and dedication by a mere percentage—or, perhaps better said, without which man's progressive spirit of adventure is consequently frustrated—would more than pay for the entire space effort.

The cons, represented by Warren Weaver, James Reston of the *New York Times,* and Philip Abelson, editor of *Science,* would argue:

1. The cost of manned lunar landing, and indeed most of space science, is exorbitant and may break us financially. If the Russians want to break themselves in this way, let them go ahead.

2. A fraction of this expenditure devoted to other aspects of scientific research would advance science much more than the space program.

3. Think of what 5 billion dollars a year would do for the abolition of poverty! Or for the advancement of health (or, more generally, we could use the money better for something else).

4. If the military aspects of space are important, leave those developments to the military forces.

Then there are the intermediate positions, typically advocated by Vannevar Bush and possibly Lee DuBridge:

1. Let us learn to walk before we run. Proceed with the program at a more modest pace over a longer time scale.

2. The risk of sending man to the moon at this time is too great and his loss on an ill-advised flight would stir great public revulsion.

3. Instrumentation in the interim could do much that man could do. Why not wait and see?

19

4. If you put all your effort on the moon, the Russians will circumvent you by choosing an alternative space target —probably Mars.

Now these are by no means all of the arguments, and for each argument there are suitable answers and counter-answers. For example, the program is damned for recurrent delays, slipping schedules, and bad management; at the same time it is defended for creation of a vast new technology which requires extensive extrapolations beyond existing experience, so that substantial errors inevitably appear at each step, to be assessed and eliminated by actual test and experiment—the historical experience with all new technologies.

When the Russians back down there is demand to slow our program; others reply that we should stand on our own feet and not allow the Russians to dictate our policies. The whole point is that the argument is necessarily based on subjective grounds, involving estimates and judgments that are assessed differently by different persons. On one point, however, there is no debate: to land a man on the moon and to return him safely—at a price—is technically possible during the next decade. The fact that the national leaders closest to the program overwhelmingly favor the present course may be automatically attributed to the natural bias arising from their association with it.

Intuitively, I have supported the present course for quite another and I believe more fundamental reason: its influence in stimulating man's spirit and raising his sights, as "a big stride toward freer and fuller rationality." Since the shepherds of Babylon, man's eyes have been turned consciously toward the heavens. Now that they are within our reach, man yearns for bold steps to press the conquest. To do less would shrivel his spirit, diminish his stature.

As space engineer Richard Porter recently observed:

> The only valid rationalization seems to be that of acquiring knowledge and understanding—fundamental knowledge and understanding about the universe in which we live, and hopefully about the origin of life in it—information which will unquestionably affect our concepts of ourselves, where we came from, and where we are going, and which will influence in unforseeable ways our politics, warfare, business, religion and art. . . .
>
> I simply cannot imagine the people of this proud country content to sit in their rocking chairs watching others take the lead in this great adventure of our time.[7]

Just as the navigators of the sixteenth century contributed to the golden age of intellectual development by extending man's horizons, the space scientists of our own age are responding to a challenge of comparable magnitude, the challenge of space. Are not man's restless energies and new technological capabilities better invested in the conquest of space than in futile wars of aggression? This view is cogently urged in the words of the Final Report of the Twentieth American Assembly,[8] a group of distinguished Americans meeting at Arden House during October 1961:

> Before long, human footsteps will imprint the dust of the moon's surface. As more and more satellites cross the skies, and as man's old dream of contact with the moon and the planets is transformed into reality, human life will inescapably take on new dimensions. Not the least of the motives impelling us will be the human and cultural values involved in pursuing the high goals of knowledge about

7. Address, Clear Point, Alabama, Oct. 10, 1963.
8. *Outer Space: Prospects for Man and Society,* ed. Lincoln P. Bloomfield, The American Assembly, Columbia University, Prentice-Hall, New York, 1962.

our origins and our destiny. Few are the generations privileged to take part in a comparable enterprise. For this adventure, and for all of its social, economic, scientific, military and political consequences, our nation can and should pledge both its generous support and its responsible leadership.

As Shakespeare sees man, "He rises on the toe. That spirit of his / In aspiration lifts him from the Earth."[9]

Out of these broader contexts, our major judgments in the application of science and technology should be formed. Although they require familiarity with the limitations of the related sciences and technologies, these are human judgments, which science, as yet, has no means of effecting. They are judgments that may determine whether our society can find the means of easing the tensions produced by the age-old gap between the have and the have-not nations— whether man can be motivated to enjoy the full fruits of our scientific age and thus to enhance his own environment.

9. Shakespeare, *Troilus and Cressida*, IV.5.15–16.

# 2. Advanced Education for a New Age

Last year, Clark Kerr,[1] Chancellor of the University of California, gave a series of lectures at a well-known college not far from New Haven entitled "The Uses of the University." In the Godkin lectures at Harvard he painted a striking picture of the wide range of functions of a great university today. Commenting on the progress of the university during the past half century from the wings to the center of the community stage, Chancellor Kerr cites the changing social forces to which the university is exposed:

> The industrial, democratic, and scientific revolutions have gradually moved in on the universities and changed them almost beyond recognition. In all of these intellectual and social revolutions, the university, as an institution, was initially more of a "stronghold of reaction" than a revolutionary force, although the ideas of its individual members have often been a stimulus to change.
>
> Change, when it has come has been initiated from the outside, or at least assisted from the outside gates.

Kerr then goes on to remark, "Few institutions are so conservative as the universities about their own affairs while their members are so liberal about the affairs of others."

1. Clark Kerr, *The Uses of the University,* Harvard University Press, Cambridge, 1963.

Let us examine the question: To what extent are American universities evolving to satisfy the social needs created by the scientific revolution, and what forces are impelling this evolution?

In the last chapter it was noted that our nation is in the midst of a social and economic revolution of a magnitude unprecedented in the history of society. This revolution is nourished by science, from which is derived an equally powerful technology, and is characterized by the rapid emergence of entirely new industries founded upon scientific innovation, which give man greater control over his environment. This new resource emerged slowly at first, producing over the past century electric power, the telegraph, telephone, railway, airplane, and motor car in primitive forms. Then by midcentury the rate of innovation had accelerated to the point where it accounts for about half our present economic activity. Phenomena related to this revolution are the population explosion resulting from the conquest of disease; the increased productivity that is multiplying our wealth at an incredible rate; access, almost overnight, to the whole of our planet for its full development and now into space as well; an almost catastrophic movement of the population from the land into a hundred or so metropolitan areas; and a steady rise of unemployment centered primarily among the uneducated and the unskilled.

It is a characteristic of this technologic society that it improves productivity and quality through automation, thus releasing manpower from the older tasks. At the same time, innovation derived from a virile science creates altogether new products and services, with new opportunity for industry and employment—products and services that further improve man's adaptation to his environment. Thus the total productivity of society is continually enlarged, providing the first real hope for the early obliteration of poverty. The stability of this economy must rest primarily upon an

24

adequate "supply of innovation" and upon a people motivated by "a sense of freer and fuller rationality" to need new products and services.[2]

When the utilization of innovation for these purposes occupies an appreciable portion of society's effort, the concept of "labor" takes on new meaning. Thus the kinds of skills required in an economy of plenty are radically different from those required in an older economy, as has been pointed out in recent reports of the President's Science Advisory Committee. In particular, a representative group of men who have access to the limits of knowledge is essential to the employment of all the others.

We are led then to examine the response of our universities to these new needs for educated men. Let us take a look at some numbers. In Figure 1 we see the gross statistics for graduates at various levels in all fields, together with graduates in engineering. In all fields we now graduate about 450 thousand at the baccalaureate level, and this total should grow to some 700 to 800 thousand by 1970 if the exponential rise continues. Such a rise would require that we roughly double our educational plant in the next seven years (which, of course, we won't since we are facing an

2. The uncritical will complain that there is a "Madison Avenue" sound to this whole idea of motivating people to buy things they don't need—like wigs ($179.00) and jewelled beer-can openers! The idea of motivation to buy is unnaturally associated in the public mind with flashy but useless gadgets. This attitude looks at only the edge of the coin, and emphasizes mere superficialities. Society must penetrate more deeply to see that the nostalgic resistance to the introduction of the electric light, telephone, railway, airplane, automobile, and a hundred other "adaptive" devices and services has simply delayed the obliteration of poverty by retaining life in less productive patterns of activity. Any free system (and many not so free) will find some demand for gadgetry and the more artificial luxuries. But the central importance of motivation toward improved adaptation is not the sheer sense of luxury that it affords but the vast improvement in the effectiveness of the individual.

25

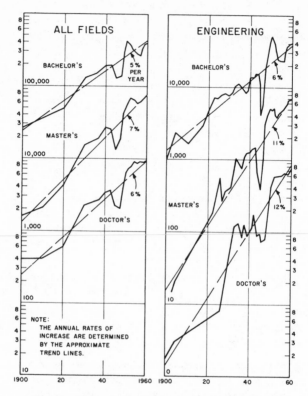

Fig. 1. Degrees Conferred in the United States
from 1900 to 1960

exponential). What is the supply? Last year our high schools
graduated some 2.8 million. By 1966, as a result of the
postwar baby crop, the number will rise to 3.8 million and
continue to rise thereafter. Taking all probabilities into account, we can estimate that by 1970 we will, *at the most,*
be graduating at the bachelor's level about the same proportion of high school graduates as at present. On the face of
it, this seems to portend real trouble in unemployment,

when our economy of plenty calls for a rapid reapportionment of skills.

Well, says the complacent educator, the rise has been about exponential over the past 40 years—doesn't that indicate good performance? It would if it could be maintained. But to maintain an exponential at the top of the curve is immeasurably more difficult than at the bottom. If you examine the curve closely you will find that it is below the exponential after 1957, and the slippage is becoming more marked. Nothing short of an extraordinary sense of urgency can maintain that exponential rate of growth at a time when society desperately needs it.

But, you say, the scientific revolution requires scientific and technological leaders. Aren't we compensating for the total deficiency by our overemphasis on science and technology? Last year we graduated some 32 thousand engineers, or about 8 per cent of the total graduating class. Engineers, together with natural scientists of all kinds, comprised 21 per cent of the graduating class. Sir John Cockroft, chairman of the distinguished education committee OECD,[3] points out that Britain, in comparison, graduates 47 per cent at the baccalaureate level in science and technology. So the common charge that we are overemphasizing science and technology in American universities at the expense of the humanities just does not seem to be supported by the facts.

Perhaps, you say, there aren't enough students interested in science and mathematics. You will see from Figure 2 that about two thirds of the baccalaureate graduands come from about fifty universities; these are also the universities that are the mainstays in the fields of science and engineering. Their classes are severely restricted. At one of these

3. Report of Committee on Advanced Education, The Organization for Economic Co-operation and Development, "America's Way to Educational Reinforcement," No. 3, March 15, 1963.

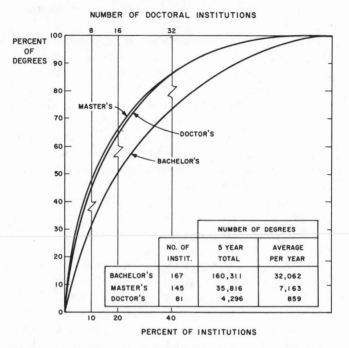

NUMBER OF DOCTORAL INSTITUTIONS

| | | NUMBER OF DEGREES | | |
|---|---|---|---|---|
| | NO. OF INSTIT. | 5 YEAR TOTAL | AVERAGE PER YEAR |
| BACHELOR'S | 167 | 160,311 | 32,062 |
| MASTER'S | 145 | 35,816 | 7,163 |
| DOCTOR'S | 81 | 4,296 | 859 |

PERCENT OF INSTITUTIONS

Fig. 2. Institutional Origin of Graduands in Higher Education

schools over 700 students matriculated in engineering last year and more than 400 were failed. Since not all, or even very many, top high school graduates are "dumb," this says immediately that (1) the university was deliberately conducting a ruthless weeding, or (2) there is no articulation whatever between high school training and university entrance requirements. In either case, the system is dreadful in its effects upon the individual and in depriving society of many persons potentially useful in science and technology. Dean William Everett, at the University of Illinois, has instituted a practice of qualifying high schools for matriculation of their graduates in the engineering school of the uni-

28

versity; to graduates from unqualified schools, the university offers remedial summer courses. But this practice is not yet general.

One by one, as we pull questions out from under the rug, we find that the answers demand a rather thorough overhauling and enlargement of university procedures to meet the needs of our new age. I hate to mention it, but in 1962 the Russians graduated about 130 thousand at the baccalaureate level in science and technology compared to our 95 thousand, and of comparable quality.[4]

Obviously, the ideas underlying the technology of the future are abstruse and highly mathematical in nature. To comprehend them, and to utilize intelligently the technology born of them, we require men of very advanced education in substantial numbers. This does not mean a mere four-year college education, for the ideas involved are at the very boundaries of knowledge. Command of the new technology and of the science from which it is derived requires postgraduate education to the doctor's degree and beyond—not less than eight years beyond the high school diploma. In particular, it may well require in the immediate future continued education at the graduate level, concurrent with productive endeavors, throughout a lifetime.

Of course, men of lesser training are essential and can be usefully employed in technological industry. But, for each Ph.D. we can employ five to ten engineers, and for each engineer we can use ten to fifteen skilled workers. The creation of new industry, new products and devices, new methods and applications from the new technology arises out of the creative and imaginative insights of scientific and technological leaders who have access to the very limits of knowledge. Without that flair for innovation at the top, men

4. Nicholas DeWitt, "Soviet Education at the Crossroads," *Bulletin of the National Association of Secondary School Principals, 47,* No. 282, April 1963.

of lesser skills will be deprived of the opportunity they would otherwise find. Therefore we cannot discuss the development of the new economy without discussing the educational needs of the men and women who must provide the top brainpower to effectuate our national research program.

I would make this point unambiguously: no training of numbers at the trade school, high school, or college level can in itself sustain the new technology. Indeed, in the future we may have to count a hundred or more unemployed for each Ph.D. we fail to educate. If, and when, we find unemployment at the intermediate level of engineering and technology, we must not be misled into imagining that we are overproducing engineering talent. Quite to the contrary, this is symptomatic of the failure to train sufficient numbers who have full access to all scientific knowledge and the means of transforming it into new technologies that could employ them and the lesser trained as well. Those communities that can produce and retain men of advanced education will have the most immediate access to and control over the new technology from which the industry and wealth of the future will flow. They will have the opportunity to provide for full employment at all levels.

This is why extensive doctoral and postdoctoral education in the sciences and engineering has suddenly become imperative. This is why the new science-oriented industry is suddenly exploding around the great centers of advanced education in Boston, New York, San Francisco and Palo Alto, southern California, and Minneapolis–St. Paul. For the science-based industry must have access, not only to highly trained men in their own organizations but also to the sources of education—the universities and the great fundamental research laboratories where the secrets of nature are being explored to their very limits.

With essentially our whole population flooding to the

hundred or more metropolitan areas, and becoming ever more dependent on science-based industry for employment, it seems clear that our economic and social health will require in each metropolitan area at least one great university competent to graduate no fewer than 200 Ph.D.s each year. Our national average should exceed 100 Ph.D.s per million people per year.

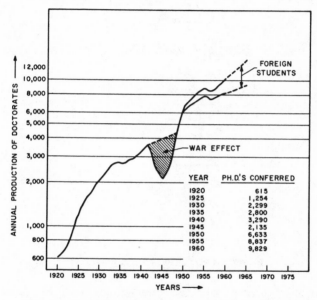

| YEAR | PH.D'S CONFERRED |
|------|------------------|
| 1920 | 615 |
| 1925 | 1,254 |
| 1930 | 2,299 |
| 1935 | 2,800 |
| 1940 | 3,290 |
| 1945 | 2,135 |
| 1950 | 6,633 |
| 1955 | 8,837 |
| 1960 | 9,829 |

Fig. 3. Doctoral Degrees Conferred by United States Universities, 1920–1962

Where does our nation stand in the face of this revolutionary social and economic change? As shown in Figure 3, in 1962 we produced a few more than 11,000 Ph.D.s of all kinds annually (excluding doctorates in law and medicine), about half of whom are graduated in science and technology.

31

Some 55 per cent of these in the 1960 era were graduated by the twenty major universities shown in Figure 4.

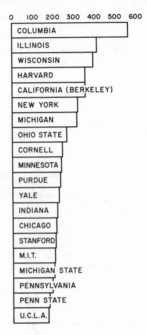

Fig. 4. Institutions Leading in Doctoral Degrees, 1960–1961

The failure of great new graduate schools to emerge in the past two decades represents a critical problem. In 1920 some ten graduate schools produced two thirds of our Ph.D. graduates. In 1940 this number had increased to twenty universities producing two thirds of the doctoral degrees. Only five universities advanced to the ranks of the top twenty between 1940 and 1960, and these were strong contenders in the 1940 era.

Now, more than two decades later, some twenty-five universities still dominate the scene and these are approaching

## TABLE 1.

*States Leading in Doctoral Education*

Ph.D.s for 1961–62
per million
state population

| | |
|---|---|
| Massachusetts | 158 |
| Colorado | 125 |
| Iowa | 122 |
| Indiana | 117 |
| Wisconsin | 112 |
| Connecticut | 103 |
| New York | 91 |
| Utah | 91 |
| Illinois | 90 |
| Rhode Island | 83 |
| Michigan | 81 |
| Minnesota | 79 |
| Oregon | 74 |
| California | 71 |

*14 states, where 17 of the big 20 universities are located,
graduated about 100 Ph.D.s (average) in 1961–62*

saturation. At the very moment when our national situation demands a radical expansion of graduate education, the development of great new graduate centers has essentially halted. As a consequence, the number of doctoral graduates has fallen decidedly below the exponential rise characteristic of the baccalaureate graduands. This trend is further accentuated by the rising proportion of graduate students of foreign origin since World War II, a rise from 5 per cent in 1950 to 15 per cent in 1960. So the number of American doctorates is still further depressed in this proportion.

Well, you say, what about universities numbers 25 to 100? Don't they have a golden opportunity to meet this challenge? They certainly do, but will they interpret the

33

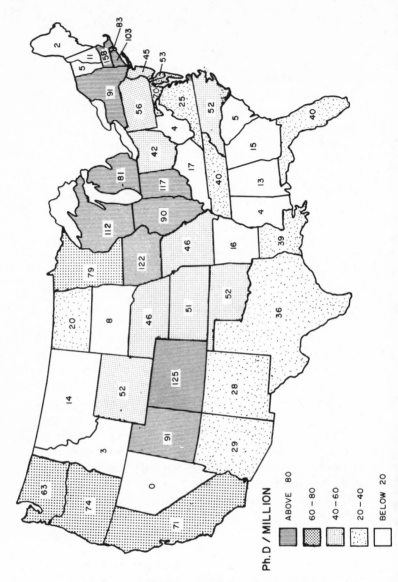

Fig. 5. Distribution in the United States of
Doctoral Degrees Conferred, 1962

needs of our age as a challenge to them? Perhaps the problem can best be seen in perspective at the doctoral level.

Each of the Big 20 is graduate-oriented, with more than 35 per cent of its students in the graduate schools. Universities ranking in size from 21 to 100 are primarily undergraduate-oriented, usually with fewer than 12 per cent of their students in graduate schools.

Fourteen education-oriented states (with 17 of the Big 20 universities) graduated an average of about 100 Ph.D.s in 1962 per million population (Table 1). Primarily, these are the states of the Northeast, the mid-North, the Mountains, and the Far West, as shown in Figure 5. The

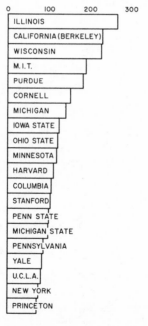

Fig. 6. Institutions Leading in Science
and Engineering, 1960–1961

remaining thirty-six states award from seventy down to zero doctoral degrees per million population per year.

Students do not undertake long journeys to attend the great universities. In 1963 fewer than 1,000 traveled more than 500 miles to earn their doctorates. Living near a great graduate school, about 12 students per 1,000 high school graduates will receive this degree and, at greater distances, fewer than 5 per 1,000 will achieve it.[5] This does not mean that a student always goes to the nearest graduate school—it does mean that the great graduate schools provide a powerful intellectual motivation in their communities.

We have spoken so far of total doctorates earned. Of particular interest is the fact that of the top twenty universities, eighteen also award most of the Ph.D.s in science and engineering, as is seen from Figure 6. Moreover, among the Big 20 more than 55 per cent of the doctoral degrees are in science and technology whereas the remainder average less than 45 per cent in these intellectual disciplines.

The contrast among metropolitan areas is even more striking (Fig. 7). Compare Boston, having 600 Ph.D.s annually, or San Francisco with 500, or Los Angeles or Detroit having 300, or Minneapolis–St. Paul having 250 with Houston, New Orleans, or Oklahoma City having 40, or Dallas–Fort Worth having 6, or Phoenix with none. Or compare a rural university like Cornell, at Ithaca, having 225, or Illinois at Urbana having 350, and the only substantial graduate university in the South or Southwest, Texas at Austin, having about 150. These contrasts may be bitter, but they are too critical to ignore in the face of the advancing technological revolution.

Without question, Americans must take off their hats to the leadership of the twenty or so great American universities which represent not just the core, but almost the

5. National Research Council Manpower Study No. 3.

## PH./D. PER MILLION (APPROXIMATE)
## 1959

100        200

MILWAUKEE-MADISON
ALBANY-SCHENECTADY-TROY
BOSTON-PROVIDENCE
INDIANAPOLIS
DAYTON-COLUMBUS
MINNEAPOLIS-ST. PAUL
DENVER
SAN FRANCISCO-BERKELEY-SAN JOSE
SEATTLE
DETROIT-ANN ARBOR
WASHINGTON-BALTIMORE
NEW YORK-N. NEW JERSEY
PHILADELPHIA
CHICAGO
KANSAS CITY
PITTSBURGH
BUFFALO-ROCHESTER
L.A.-LONG BEACH-SAN BERNARDINO
CLEVELAND
NEW ORLEANS
ST. LOUIS
HOUSTON
ATLANTA
CINCINNATI
BIRMINGHAM
LOUISVILLE
PHOENIX
DALLAS-FT. WORTH
PORTLAND

METROPOLITAN AREAS

MEMPHIS-O            SAN DIEGO-O
        SAN ANTONIO-O
MIAMI-O              TAMPA-O

Fig. 7. Distribution by Metropolitan Areas of
Doctoral Degrees Conferred, 1959

whole of significant American graduate effort. To these we must add a few smaller universities such as Caltech, Rice, and Princeton, whose unquestioned excellence by any standard of international comparison has contributed immeasurably to the quality of our doctoral training.

But at the same time we have the right to ask about universities 25 to 100. I remind you of Clark Kerr's implication that universities like solid bodies behave according to Newton's first law, remaining at rest or in a state of oblivion until pushed by an external force. In other words, the intellectual leadership responsible for the welfare of every metropolitan area must come alive and do some pushing. Some communities are already awakening, and during the next decade it will be interesting to observe the results.

Well, you may say, do you propose to generate new institutions that are mere Ph.D. factories? I remind you that the numbers we are considering are relatively small. Our 1960 output of doctorates was only 650 in physics, 800 in engineering, 1,200 in chemistry, 1,000 in biology, and a handful in mathematics.

On the other hand we can reliably estimate that more than 75 thousand high school graduates annually *exceed* the median level IQ and creativity indices for individuals now receiving the Ph.D.[6] The median for physics is only 130, and for other disciplines only slightly below this; these indices are exceeded only by the Ph.D.s in psychology at 138, and psychologists write the tests! It seems obvious that we are utilizing scarcely 5 or 10 per cent of the potential supply. How greatly the caliber of our Ph.D. crop would be improved if we could but capture some of that highly qualified 90 per cent before they are siphoned off into attractive but ultimately less productive pursuits. This will require a mighty reorientation of universities 25 to 100, immense

6. See Lindsey R. Harmon, "High School Backgrounds of Science Doctorates," *Science, 133* (March 10, 1961), 686, Table 2.

expansion of facilities, and better articulation of the student from high school to the university.

Or you may ask whether graduate training should be tied to frankly economic or military objectives. Should not the case for advanced education be made in terms of purely scholarly goals and man's creative and spiritual growth? Are you not degrading academic aspirations by relating advanced study to man's material welfare? Are you not stressing more training to acquire a salable skill in contrast to education which "cultivates reasoning ability, creativity, tolerance, eagerness for new ideas, a sense of history and of potentialities for the future"?[7]

No one can question either the intellectual search for truth as a major objective in the advance of the human spirit or man's striving for the expression of beauty, compassion, and dignity. Certainly the foundation of knowledge must always have as a major objective man's apprehension of his total environment—his entire adaptation to it. But as we have seen, our society at midcentury finds itself plunged into a new situation to which it must suddenly adjust by a far greater emphasis on advanced education.

In this changed situation a variety of new factors and forces dictates a critical reappraisal of some traditional clichés in higher education. The decline of liberal arts in their Victorian form is mourned by such humanists as Jacques Barzun[8] who sees them "being invaded, not to say dispossessed, by the advance agents of the professions, by men who want to seize upon the young recruit as soon as may be and train him in a 'tangible skill.' "

The decline of the liberal arts is indeed a source of concern to us at a time when man has great need for all the inner strength that they can provide. But I fear, Barzun to the contrary, that the invasion of the professions has oc-

7. Jacques Barzun, address to Hofstra University, Dec. 12, 1963.
8. Ibid.

curred, not so much out of deliberate dispossession, but rather because the liberal arts have somehow failed to grasp and assimilate the great seminal ideas of modern science. Originally, of course, there was no dichotomy, and science gradually found its way into the curricula of the universities bearing the label "natural philosophy." But as the applications of science began to demonstrate their usefulness to society, science and engineering tended to go one way and the liberal arts another.

This is a great pity because the educated person, regardless of his professional specialization, needs to have an appreciation, at least, of both kinds of knowledge. Crane Brinton[9] has conveniently labeled these *cumulative knowledge,* illustrated by the sciences, and *non-cumulative knowledge,* illustrated by literature and the arts. He notes that:

> A modern American college student is not wiser than one of the sages of antiquity, has no better taste than an artist of antiquity, but he knows a lot more physics than the greatest Greek scientist ever knew. He knows more *facts* about literature and philosophy than the wisest Greek of 400 B.C. could know; but in physics he not only knows more facts—he understands the relations between facts, that is, the theories and the laws.

Brinton observes further that:

> The intellectual historian clearly must concern himself with *both* cumulative and non-cumulative knowledge, and must do his best to distinguish one kind of knowledge from another, to trace their mutual relations, and to study their effect on human behavior. Both kinds of knowledge are important, and each does his own work on this earth.

A major problem certainly, is the vast accretions of knowledge in both categories. If Bacon[10] was able to say,

9. *Ideas and Men,* Prentice-Hall, New York, 1950, pp. 13–15.
10. Letter to Lord Burghley, 1592.

at the very dawn of modern science, "I have taken all knowledge to be my province," the great Newton[11] was saying less than a century later:

> I do not know what I may appear to the world, but to myself I seem to have been only like a boy playing on the seashore and diverting myself in now and then finding a smoother pebble or a prettier shell than ordinary, whilst the great ocean of truth lay all undiscovered before me.

The sheer volume of knowledge, when taught in the historical order of its discovery and formulation, has gone far beyond the capacity of the individual to absorb in an ordinary university career. Taught in this way, science and mathematics seem difficult and abstruse. Consequently, Brand Blanshard,[12] Professor Emeritus at Yale, is led to comment that "science is a bore," though I suspect he means that science as now taught is a bore. Judging from John Fischer's editorial comment in a recent *Harper's*,[13] the same can be said about the teaching of English; and one further suspects that this generalization can be extended to the current teaching of most branches of knowledge.

The difficulty, arising from the unsuitable organization of knowledge as now taught in monotonous historical sequence, raises a formidable barrier between the sciences, and the now equally specialized liberal arts, for those who genuinely desire to feel at home among all of the great issues of human thought. The real hope of surmounting this barrier lies in the continual reorientation of teaching around the great and exciting central ideas and generalizations in

11. David Brewster, *Memoirs of the Life, Writings, and Discoveries of Sir Isaac Newton, 2,* Chap. 27, Constable, Edinburgh, 1855.

12. Symposium: Communication between the Arts and Sciences, Oct. 27–28, 1961, Kenyon College.

13. Feb. 1964.

each area, leaving the details, the grammar and syntax, the axioms and special propositions, the ultimate limitations of generality, and the historical perspectives as interesting and useful avenues to be pursued only after the central notions have been rather fully digested. Some progress, at least, is possible so long as there is an awareness, in both the sciences and the liberal arts, that each has something to contribute to the other and that the universities have a responsibility to find feasible ways in which to reunite these two great streams of learning.

In considering the urgent need for education to keep abreast of the needs of contemporary society, we might well recall the lectures of that remarkable pedagogue, a certain Professor Peddiwell, whose unusual and entertaining analysis of Paleolithic education was stimulated by a liberal supply of tequila "daisies" during the racing season at Tia Juana, and later published under the title of *The Sabre Tooth Curriculum*.[14] It seems that in the early days of Paleolithic tribal culture, the tribe subsisted by catching fish with their hands and clubbing horses for their hides. Since the tribe was plagued with poverty, a great leader instituted a university to teach "fish-catching with the hands," and "horse-clubbing" as well as "tiger-scaring with torches," for fierce sabre-toothed tigers barred the way to the tribal waterholes.

With these sophisticated and educated means of meeting essential needs, the tribe flourished. The three central courses of the university were subsequently elaborated to develop in detail every intellectual facet of these basic themes. In particular, the courses on tiger-scaring culminated in a stirring ceremony in which the butt of the torch was kissed with great reverence.

14. J. Abner Peddiwell, with Foreword by Harold Benjamin, McGraw-Hill, New York and London, 1939.

As time passed, the great glaciers approached, completely changing the tribal environment. The tigers migrated to more favorable climes, and with muddy streams and fleeter horses, the tribe turned to a more advanced technology for their subsistence. Yet the university continued to preserve the traditional curriculum of fish-catching, horse-clubbing, and tiger-scaring, and refused to permit the introduction of new courses based on the innovations to which the tribe had turned for its subsistence and welfare. For how, reasoned the faculty, can our students acquire a sense of nobility and courage, or the ability to think or reason clearly without the opportunity to "kiss the butt of the torch"?

Let us move on to other aspects of science education. At doctoral and postdoctoral levels scientific research is an indispensable part of the training process. Graduate education and scientific research belong together for the following reasons:

1. Graduate teaching is sterile unless the faculty actively participates in research. The whole attitude of the student is different under the intellectual leader whose thinking is oriented by his search for knowledge. Without research, teaching at the graduate levels becomes stereotyped and quickly outmoded. As new graduate institutions are established, research opportunity must be extended to the faculties of all of them.

2. Graduate students must be trained in the methods and procedures of scientific and engineering research by actual participation under skilled and experienced investigators. Scientific proficiency requires the ability to engage in the experimental manipulative processes of research.

3. Some of the most significant new insights into scientific problems come from fresh new minds in their early contact with the problems that nature presents.

But quite beyond these educational needs is the experience of the past two decades that the research-minded faculty becomes a tower of strength to society—both to government and to industry. Scientific knowledge is almost useless and soon incomprehensible to a people when it is merely stored in books. Science is something alive—it consists of people with the basic knowledge and the education to use it, with skill in tackling novel situations, who know a thousand and one ways of circumventing difficulty in developing experiment and elaborating hypothesis, and who have a highly developed and constantly employed creative experience.

A scientific tradition can be a live tradition only as it is perpetuated by the never-ending transmission of the whole body of knowledge and technique from master to student. The university combines the closely related functions of teaching and research by producing better students and advancing knowledge in a single operation. The university is the natural home for research-minded men from whom the community draws its strength in a variety of ways.

Since the new society will require at least one great graduate institution in each of the hundred metropolitan cores of our nation, we face the problem of developing new institutions at the true graduate-university level in each core where one is now lacking. The failure to develop such institutions after the 1930 era stems from a variety of forces, all acting to discourage such ventures. Among these forces are:

1. The absence of powerful leadership from the liberal arts.

2. A tendency to cling to the "gentlemanly" curricula of the Victorian era.

3. The illusory but widespread tendency to imagine a university as a group of massive buildings, or as a faceless curriculum with a multiplicity of courses, rather than a

scholarly faculty which, with the students, forms the university with the external trappings as useful adjuncts.

4. The ridiculously low pay of faculty and outrageously high teaching loads that leave little opportunity for individual faculty development or student guidance.

5. The subordination of scholarship to athletics and other extracurricular objectives, possibly as an emotional escape from intellectual frustrations.[15]

6. The "rah-rah" alumni body and its preoccupation with football, homecoming, and campus beauty queens, who are mentally still in the 1920s.

7. The haphazard introduction of specialized professional training in the freshman year as a consequence of the failure of the liberal arts to give the student an appropriate introduction.

All of this has combined to produce a comfortable mediocrity, well insulated from the major social revolution of today, largely incapable of comprehending it and certainly not of exercising leadership over it. Only the pressure of well-informed public opinion, applied in suitable directions, can create new visions of scholarship, new standards of social value, new responses to the vitality of society itself.

Yet unhappy as this picture appears, I don't think it proper to assess the full blame on the universities that have failed to make the grade. Many are trying hard, desperately hard, to rise above this dismal status. The very existence of the top 20 or 25 makes it well-nigh impossible to attract

15. Realistically, taking into account the relative pay in the athletic department, and of the remaining faculty, as well as the relative value of scholarships, many such institutions should be characterized primarily as schools of physical culture with a few poorly paid professors attached to add an air of respectability. Good professors are as expensive as good coaches, and both have a place in a properly balanced college or university.

scholars to what should be new centers of learning. Thus, rising to graduate status today, in the face of the distinction of the Yales, Harvards, and Princetons, the MITs, or the Berkeleys and Wisconsins, is a completely different problem than those universities had to cope with in their adolescence.

TABLE 2. *Date of Founding of Principal Southwest Colleges and Universities*

| | | | |
|---|---|---|---|
| Tulane | 1834 | New Mexico State | 1889 |
| Southwestern (Georgetown) | 1840 | New Mexico Inst. Mining/Tech. | 1889 |
| Baylor | 1845 | Oklahoma | 1890 |
| Austin (Sherman) | 1849 | Oklahoma State | 1890 |
| Louisiana State | 1860 | North Texas State | 1890 |
| Trinity | 1869 | Louisiana Polytechnic | 1894 |
| Arkansas | 1871 | Southwest Louisiana | 1898 |
| Texas A. & M. | 1871 | Texas Woman's | 1903 |
| Texas Christian | 1873 | Rice | 1912 |
| Texas | 1884 | Southern Methodist | 1915 |
| Arizona | 1885 | Texas Tech. | 1923 |
| Arizona State | 1885 | Houston | 1934 |
| New Mexico | 1889 | Dallas | 1956 |

In this context, I should like to describe for you a regional experiment directed explicitly toward filling this hiatus in one great region of our country, the Southwest, which represents one of our last frontiers of natural development. Over the past century it has become a major agricultural region producing cotton, cattle, grains, fruits, and vegetables. It has oil and minerals in profusion; its rich natural resources have made it a land of opportunity that has attracted population in great numbers. Its development on the very threshold of the American "Great West" is the theme of many a song and tale.

During these years, more than one hundred colleges and

universities have been established in the region, primarily for baccalaureate training. Because the region has been so recently developed, only five of these institutions are as much as a century old (Table 2). The rapid growth of Southwest industry and education during the past several decades magnifies the problem of grasping the opportunities made available by science. Frontier patterns have been abruptly changed by technological advances. Almost overnight, the Southwest has developed a critical need for greatly expanded graduate programs that will permit it to take advantage of the opportunities that are there today.

How can a great people meet the challenge to increase substantially and quickly the number of highly trained minds? Clearly the only source can be graduate schools in our universities. Which universities should accept the challenge? How many such institutions does the regional community need, and should it support? How can existing undergraduate institutions rise to graduate stature in a tradition of excellence? How can existing postgraduate programs be expanded and made even better?

These were the questions troubling southwestern leaders when I first became aware of this problem. One evening in 1958 on a flight from Dallas to New York, Erik Jonsson, Chairman of the Board of Texas Instruments, Incorporated, of Dallas (and now Chairman of the Board of the Graduate Research Center of the Southwest) opened my eyes to the social and economic revolution into which the world was plunging. He described the picture in terms of the problems that our newly developing technological revolution presented at the recent frontiers. I recalled my earlier talks with Dean McGee, President of Kerr-McGee Industries, Incorporated, of Oklahoma City, who likewise had emphasized the quickly changing social and economic patterns of the region. These were men of immense stature, who through experience and foresight were deeply conscious of the future and of the new resources required to carry their region beyond its current levels of achievement.

The solution to the problems, in the minds of these men (and as I was soon to find, in the minds of many, many others of the Southwest) hinged on the rapid development of graduate education (see Fig. 8)—the need in every community for minds developed to deal with novel situations. To them, the whole health and happiness of the future lay in a quick and intelligent response to the new problems of our technological world.

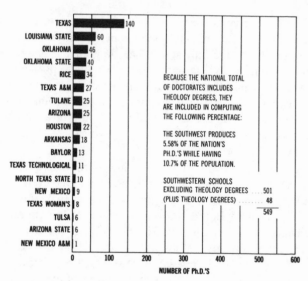

Fig. 8. Doctoral Degrees Conferred by Southwestern Universities, 1959–1960

In the ensuing months, in an attempt to understand the fundamentals of the situation,[16] I developed a series of

16. Subsequently published under the title, *Graduate Education in the Southwest,* SMU Press, May 1961.

studies of the Southwest, which grew out of conversations with many leaders who were conscious of the real meaning of events of today.[17] The proposals in these studies were vigorously debated one historic evening (May 27, 1960) in the rooms of Southern Methodist University, with a group of southwestern leaders.

The essence of the proposals was that a regional institution dedicated solely to high scholarship should be founded at Dallas, with the objective and ideal of scholarly research and postdoctoral education, to aid in the development of a regional program for academic excellence. The new institution would work with existing universities in the region in building graduate opportunities as rapidly as possible. It would work with industry as a source and storehouse of fundamental thought out of which innovation would be encouraged.

The institution would devote itself entirely to basic research and advanced teaching and would not be in competition with the regional universities in their predoctoral work. It would be organized as a university with the highest standards for faculty selection; it would be qualified to grant degrees to ensure its academic stature, but it would not do so in practice, in order to encourage university development throughout the region. In its initial phases, it would devote its work primarily to the natural and life sciences, using them in concept as the sharp spearhead for one of the great experiments in social science of modern time. It would become a true "community of scholars," to aid in closing the hiatus at the graduate level.

The basic plan was to work from the top down. Only when real scholarship at the advanced level was broadly developed in the Southwest could the full potential of the

17. Cf. "Whither Graduate Education?" *Physics Today,* July 1963; "The Technological Revolution of Today," *Journal of the Franklin Institute,* June 1963.

universities—and indeed the whole region—be realized. Such an institution could earn and enjoy an international reputation. The experience of such centers as the Brookhaven National Laboratory had shown that their direct influence could extend for many hundreds of miles throughout the region.

In dedicating its efforts to the "development of the mind to cope with novel situations," the planning group recognized that the Center could initiate various aspects of applied research; but, with unanimity, the group felt that research directed toward particular products and services is properly a function of industry. Therefore, the Center would focus solely on basic research. However, to accomplish well this task of technological innovation, industry would have to have intimate access to the whole substratum of an advancing science.

Thus the task of the Center and of the universities of the region was, and is, to create a sense of intimate contact between science and industry, based on the broad scientific foundations that advanced research and education could provide. The Center was to aspire to the creation of an intellectual environment within which every element of the region could flourish. If successful, the Center would be able to measure its achievements within a decade and a half.

The May evening meeting ended with the clear determination to undertake the founding of such an institution. In the following December I joined with the group to bring that institution to life. In the two years since its founding, the Graduate Research Center of the Southwest, with its academic arm, the Southwest Center for Advanced Studies, has brought together about 150 faculty and staff members. The number will grow to some 1,000 in the next half decade. Already a wide variety of cooperative teaching and student programs has been initiated with institutions of the area.

The faculty of the Southwest Center for Advanced Studies

is organized in a series of scientific centers or laboratories cutting across the lines of university departmental organization and focusing on interdisciplinary scientific problems. A growing number of faculty from the universities of the region and from the Center hold joint appointments so that they function both within departmental lines in the university and in multidisciplinary lines within the Center. In order to avoid mere transfer of faculty from one institution of the region to another, we do not recruit faculty from southwestern institutions except for joint appointments.

The number of postdoctoral research associates is expanding rapidly as the faculty grows and soon will pass the hundred mark. Substantial numbers of graduate students, matriculated in graduate schools of the regional universities, are working at the Center on dissertations through arrangement for exchange and adjunct professorships. Joint graduate programs have been created and more are projected. For example, the joint doctoral programs in geophysics created by Southern Methodist University represent one of the finest faculties in this field assembled anywhere. The Graduate Research Center of the Southwest is a unique institution, not patterned after any other, but using the integrated experience of many, in directing its efforts to a major goal—meeting the need for a greatly expanded graduate education effort in the Southwest.

With an increasing concern for the role of higher education in the new society, Governor John Connally of Texas, under an act of the legislature, has established a special committee of citizens who are studying the adequacy of education at every level in that state. In still other directions, Southern Methodist University at Dallas, as a typical example, has completed a thorough internal study and, as a consequence, is planning an extensive reorganization of its resources in recognition of today's needs. But above all, the transcendent goal of high scholarship has been accepted

51

by the community as the sole and proper institutional objective. These are significant responses to the social revolution of our time. Other responses must occur in forms suitable to each of our core metropolitan areas.

In the first chapter we raised a question about the stability of the new society of plenty. Now we can identify education as a critical factor in that stability. Apparently a modest level of intellectual achievement will foster automation, and with it an ever increasing and devastating imbalance in employment. Only a well-balanced educational pattern with emphasis on graduate attainment has the hope of fully exploiting our new resource—innovation.

Any reduction in education and related scientific research below some viable level or serious delay in extending it to the remaining half of our population could completely upset the economy of plenty by threatening its essential requirement: sufficient brainpower and a balanced distribution of that brainpower. The decline of liberal education enhances the danger of this unstabilizing influence through its failure to equip the leaders of our society with the means of exercising critical judgment with respect to our most pressing problems. In the absence of a vital, comprehensive, and dynamic humanism, the new society is rising without that broad guidance which it has the right to expect.

# 3. Science and Government

The experience of World War II demonstrated vividly to the world that science and technology had come of age as a major influence in society. For the first time in man's history a war had been fought, and won, largely with weapons that had been beyond general conception at its beginnings five years before. These new weapons were not the product of invention. They grew out of a synthesis of the most advanced ideas of our science—wave mechanics, electromagnetic theory, physical chemistry, cybernetics.

From this dramatic demonstration the scientific leader of our war effort, Vannevar Bush, produced his prophetic volume, *Science, The Endless Frontier*.[1] Recognizing that federal support of science during the war had produced a military revolution in a mere six years, Bush foresaw that science, with adequate federal support, would bring about an unparalleled economic and intellectual revolution in the postwar period.

Since the turn of the century, industrial research had grown steadily, motivated by profits deriving from the discovery and development of useful products and services. But

1. A Report to the President. U.S. Government Printing Office, Washington, 1945.

as the cream of the more obvious invention was skimmed, the ever lengthening economic cycle from the more abstruse and ever more extended research to the ultimate product was becoming too long for the risk of conventional capital. Moreover, because of the element of serendipity in the most basic scientific research, the emergent applications did not always meet the strategic objectives of the sponsoring company. It was not necessarily profitable, therefore, to project a large industrial program of basic science. So forward-looking industry began to search for ways to link the results of basic research done in the universities and other nonprofit institutions with the industrial know-how for creating new products and services.

The war showed clearly that federal support of scientific research could vastly enlarge the threshold of ideas from which a wide variety of new technologies could be drawn. Under the initial stimulus of the military establishment, and particularly of the Office of Naval Research, the federal support of scientific research that had been initiated during World War II was broadly extended. Although the national defense and welfare furnished the obvious motivation for such support, the concomitant benefits of new basic knowledge may ultimately prove to be of greatest worth to our society.

To maintain and to enlarge this growing scientific tradition, the great federal granting agencies—the National Science Foundation, the National Institutes of Health, the Atomic Energy Commission, and the National Aeronautics and Space Administration—have been created within the structure of our Federal Government. Without question, the effectiveness of these agencies has been phenomenal. No other nation enjoys equal scientific and technological support. Indeed our transition from an economy of scarcity to an economy of plenty is directly traceable to the prescience of Vannevar Bush and his colleagues in actively demonstrating the need for adequate assistance to science and advocating its fulfillment. Credit is also due to the political leaders who created the enabling legislation.

The foresight of a nation in embarking on a comprehensive program of fundamental scientific research on a large scale is worthy of special comment. It is not at all obvious to the mainstream of a society that the exploration of the curiosities of nature is likely to change the whole basis of its economy. Indeed, other nations such as Britain and Germany had also observed the consequences of scientific research during the war, yet they were not impelled to launch research programs on a comparable scale. Only the United States and the Union of Soviet Socialist Republics proceeded deliberately to develop large-scale scientific research as a major national resource.

After particular characteristics of nature have been explored, and their behavior understood, it is relatively easy to foresee the technologies that can be developed. Indeed, as we shall see, such expansion occupies 90 per cent of our present total research and development effort. But it is much less obvious, it requires much more sophistication to understand, why a nation should support scientists who ask the really fundamental questions about nature, questions that appear to lead to no obvious application, questions that may seem trivial or even foolish to the uninitiated. Even scientists and engineers, who know that a certain proportion of the answers is bound to pay off, are often surprised when they do, and are often even more surprised when the payoff comes from some seemingly unrelated and unanticipated research area. After all, man had known for thousands of years that glass would attract objects when rubbed with silk, or that the lodestone would attract other ferrous bits, or that a frog's leg would kick when connected by wire to pieces of zinc and copper immersed in a saline solution. Yet our industries of electric power, communications, and electronics waited on investigation of these trivialities, which culminated in the systematic researches of Cavendish, Faraday, and Henry. No one can foresee which properties of nature, when methodically studied, can be used effectively by technology to adapt man better to his environment. But we can be sure that some of them can and will. So as a nation we have undertaken such systematic study on a broad scale.

Even more difficult to comprehend is the idea that a new technology rarely emerges from a single scientific discovery. Technologies represent the synthesis of hundreds or thousands of discrete scientific advances, which in turn depend upon a wide variety of precedent scientific steps. Thus, to create a Mach 3 supersonic aircraft depends upon innumerable scientific investigations, yet to create such an aircraft is simply the extension of an already well-developed technology. When we can discover the controls that govern the flight of migratory birds or the return of eels to the breeding ground of their forebears, or how nature exercises its chemical control of replication of living entities, we are likely to create technologies far more powerful than man has yet seen, which will likely come from a variety of systematic studies in completely unexpected directions. The great names associated with scientific discovery are often but the tip of a whole pyramid of previously disconnected researches whose synthesis suddenly discloses the central scientific truth.

Each new technology derived from science has a permanence that continues to benefit society indefinitely in the future. Thus the capital represented by discovery outlives all other forms. Consequently, the investment in basic research should be written off over an indefinitely long time against the permanent gains acquired by society.

These concepts are not at all obvious. For a great nation to grasp them and to employ them as a part of its central purpose is a remarkable social phenomenon representing a high level of social maturation. And the result, in creating the economy of plenty, has been equally remarkable in justifying the foresight of the leaders who advocated this course. Since in ignorance it is easy for a public to assume that our progress "just happened," it is vital that the new society be keenly conscious of the deliberate planning by our leaders in creating the new economy and of the continued measures required to maintain its health. So we shall turn to the rise of support for research and development since World War II and the new social problems that are engendered.

The growth of that support is shown in Figure 9. We must recognize that about 10 per cent of the total of 15 billion dollars shown here is allocated to basic science, of

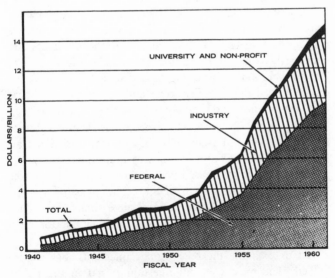

Fig. 9. Growth of Research and Development Funds, 1941–1961

which less than one billion goes to support science in the universities and related institutions. We must observe too that the contribution of the universities is probably considerably underestimated, since much of their support of graduate faculties should properly be credited as a research expenditure. Yet the total expended for really basic research to explore entirely new properties of nature is relatively small, representing less than 0.2 per cent of our gross national product. Nevertheless, the support now derived from federal sources has become a major source of funds for the development of our leading graduate institutions. Consequently, a whole series of questions arises concerning the allocation and administration of such large resources.

On one hand, the universities feel the influence of federal administration on the development of their character and on the nature of their growth. The form of all law and regulation inevitably shapes the character of the society and its institutions. On the other hand, in making large public grants for science, the Executive and the Congress are necessarily concerned that the funds be used effectively. They ask how the merits of a particular research project can be judged and what research should be supported. These questions do not lend themselves to simple answers.

This year the Congress established in the House of Representatives, under the chairmanship of Congressman Carl Elliott, the Select Committee on Research to examine these complex questions. Since science plays such a direct role in the support and growth of the economy of plenty, the emergent policies will determine whether that economy remains stable or whether it will eventually falter for want of the underlying science to support it.

Let us examine the basic policy questions that are involved. The most frequent question raised in the public mind is whether federal funds support unnecessary duplication of scientific effort. I recall that in the late 1920s, when

I followed the ideas of Merle Tuve at the Carnegie Institution of Washington in sending radio pulses to explore and identify the ionized state of our outer atmosphere, I had no idea that these methods would become the basis of a new multibillion dollar industry—radar; or the more specialized military weapon—the proximity fuse. Yet from these researches during the ensuing decade emerged a whole series of defense weapons and methods and the means of safe flight and control of aircraft and rockets—indeed the means of sight and control at great distances and in all weather.

Moreover, the duplication of these researches, each from a slightly different point of view, at the National Bureau of Standards, the Naval Research Laboratory, the Bell Telephone Laboratories, and at institutions in Britain, France, and Germany, greatly stimulated and sharpened our thinking. Thus the concept of radar steadily emerged, and when the war came it was available as a new and powerful device. But it would have been very easy for someone to kill that research in the late twenties or early thirties as "a lot of hooey and a waste of dough." Since our hindsight is always 20–20, how can we develop an equal foresight in our sponsorship of research?

A series of critical guides has, on the whole, been rather meticulously followed by the granting agencies in selecting suitable problems for support. Basically, the federal government must ask what criteria should be adopted to determine whether a proposal meets the qualifications for support. In my opinion these criteria are five:

1. Does the basic research objective proposed by the investigator pose an intelligible question for which a scientific answer can reasonably be anticipated?

2. Are the qualifications of the investigator such that he can reasonably be expected to obtain a useful answer to this question?

3. Is the investigator working in an institution where he can obtain the resources and the assistance necessary to conduct the research under the proposal?

4. Is the proposal unique in avoiding duplication of work already under way, or is such duplication justifiable to provide alternative insights which could further the relevant field of science, or provide training of students and younger scientists, or justifiably enlarge the nation's scientific base?

5. Viewed in the broad context of the responsibility of each federal agency concerned with the support of basic research to sustain a coherent body of science in which there are no glaring gaps of uninvestigated and important questions, how does the proposal compare with other projects whose support would broaden the national base of research in the relevant area? (This point relates to the responsibility of an agency to support a variety of projects which taken together will give our national program a reasonable balance.)

In applying these criteria, we should observe first that any problem of nature about which an intelligent question can be asked in a scientific sense is worthy of pursuit. The words "intelligent question" and "scientific sense" are used advisedly because superficialities have no place in the infinite realm of the unknown. The problem must be based on a substantial idea that has some clues for exploration. The answer may lead nowhere, and that result must be accepted just like a failure in business. But unless you make some tries, just as in business, you have no successes.

Second, to pursue a problem, the investigator must be reasonably qualified to organize the attack and to recognize the implications of his results. Here, however, arises the great danger of research sterility through "orthodoxy." For example, a geologist may be convinced that only a trained geologist can contribute to his subject when, as a matter of fact, the physicist or the chemist may provide powerful new insights by bringing to bear a wholly new point of view on a problem that has resisted conventional approaches. Thus a good scientific training, enthusiasm and dedication, and an eventual demonstration of research performance, are the best measures of qualification of a scientist.

Third, some duplication is justified, certainly in the course of graduate education, but more especially in the investigation of difficult problems. All great scientific discoveries have come as a consequence of the interaction of several great minds viewing a problem with different insights and skills. To eliminate duplication would be fatal—the problem is to develop the delicate sense of how much duplication is enough and, within this limit, how expenditures for facilities can lead to optimum results.

Finally, each federal agency has a responsibility to view the national research program as a whole. There is always a tendency to overemphasize the areas of science in which we have our greatest skills and to neglect others as they come into view. Unfortunately, there are large areas of sciences, some under active exploration abroad, *for which a student can find no opportunity for leadership in any American university.* Our research as a whole suffers for want of expertise in such neglected fields, which often restricts progress in related studies. Thus today one must go abroad to find skills in the important field of gas-reaction kinetics which has a powerful influence on the advancement of other aspects of science. Many other examples could be cited.

The more fundamental the problem, the greater the reward to society for a successful solution, but also the greater the chances of failure. The possibility of failure sometimes prevents the granting agency from supporting the really great problems of science. For example, as a result of recent work on the general theory of relativity, it is now possible for the first time to ask some intelligent questions about the nature of gravitation. Yet support for such research has been difficult to obtain because of the great probability of failure. Moreover, such questions are not at the moment in the mainstream of science. Discovery here might have an enormous payoff for society. It is easier for granting agencies to support less general problems whose outcome can be more clearly estimated. This attitude can denigrate real quality in our national program.

As B. D. Thomas,[2] President of the Battelle Memorial Institute, recently remarked, "If success is certain, there is no point to the experiment. Success often means the end of thought; failure may represent a fair beginning. . . . It is also dangerous to condemn research because of its apparent triviality or lack of apparent usefulness."

Thus the granting institutions must be especially sensitive to the overall balance of the program, particularly in areas where American expertise is limited. Reasonable proposals in neglected areas of science must be given some measure of priority so that students and indeed science as a whole have reasonable access to the whole range of scientific opportunity. Here there are several difficulties to overcome:

1. In new areas of scientific activity there are few protagonists in the beginning. Therefore, the ear of the granting agencies must be especially sensitive to such proposals, and imaginative foresight must be applied to ensure that promising avenues are not closed before they have been explored.

2. The granting agencies have well-developed offices to study proposals in recognized and established areas of scientific activity. There are no special offices or scientific officials to judge proposals lying outside the mainstream of current scientific effort, and they are likely to be shuttled from one office to another without action. Or if an office

2. *Science, 142* (Dec. 13, 1963), 1443.

undertakes cognizance of such a proposal, its unfamiliarity often leads to hesitancy in offering support for a project that may in fact be highly meritorious.

3. In budgeting for the Congress, allocations are established for recognized activities, with little undesignated money for scientific exploration outside the current mainstream.

4. Since the initial investigations in a "far-out" field are difficult to identify with agency or departmental objectives, the agencies fear criticism for support of proposals that cannot be clearly identified with their responsibilities or "missions."

For all of these reasons it is quite difficult to initiate scientific research involving unfamiliar scientific questions. Yet encouragement of such studies can have powerful implications for the virility of our scientific posture. "The essense of balance is to match support with the intellectual creativity of subject fields; with the need for skills at the highest level; and with the kinds of service that society currently most requires."[3]

Here the best defense for science lies in diversity in the sources of federal support. Rigidity in one agency is in part compensated by understanding and imagination in another. Nevertheless, this problem will always haunt us; the wise allocation of research money requires the most careful administration by the federal agencies.

In applying the five criteria cited earlier, the granting agencies must devise procedures that will permit the most effective program of support. Certainly an agency cannot escape the need for competent judgment within its own staff, but because of the varied ramifications of science, no staff can completely encompass the total expertise required.

3. Clark Kerr, *The Uses of the University*, Harvard University Press, 1963.

Therefore, agencies often call in panels of consultants to advise on the relative merits of proposals. Indeed, in some agencies, the panels of experts must approve a project and give it a relative rating before it can be considered for funding.

Since extensive research today is almost impossible without federal funds, the panels of experts may themselves be enjoying federal support, and to find experts who are not may be impossible. Thus the use of panels leads to certain problems. For example, the opportunity to act favorably on the project of a scientist, who may later be called upon to act on a project of your own, inevitably raises the question of conflict of interest. Consequently, charges of "backscratching" have been heard. It has even been said that only those within "the restricted circle" get the grants. A problem of another sort occurs when a panel of experts, whose selection is too narrowly limited to a highly specialized field, rejects projects out of a sense of orthodoxy. Finally, no specialized panel is qualified to judge a proposal with respect to the broad national balance of our scientific effort.

These are intensely human problems for which more adequate solutions must be worked out. First of all, the inherent integrity of the scientific philosophy offers a major safeguard. Second, panels can be more imaginatively selected to provide a broader perspective, thus avoiding orthodoxy and the cross-approval projects, even though the proprieties are nominally observed by the withdrawal from the meeting of a panel member whose project is being considered. Third, the recommendation of the panel should be just that; final judgment, either favorable or unfavorable, should be left to the agency staff which must be responsible for application of all the criteria. Finally, the agency staff should be free to employ individual, highly qualified experts to referee projects and to aid in their judgment. Actually panels tend to degenerate, becoming as conservative as their most conservative member. Sound imaginative individuals may often render the most balanced judgment.

In considering these troublesome problems we must not put the cart before the horse. Panel members are distinguished scientists whose own work contributes signally to the advancement of society and would usually therefore deserve funding by any standards. Although the application of the criteria is difficult, it must be viewed in the perspective of our broad goals, and its troublesome aspects should not be permitted to destroy the support for science and with it the basis of our new economy; instead, we should vigorously attack the problems themselves.

After selecting research problems for support, largely for the benefit of the universities, the granting agencies face a number of other serious policy problems that sharply affect the universities. The first and most important is how the federal research grant program can be administered to get universities 25 to 100 off the ground in the graduate sense. As we have already seen, there is urgent need for at least one substantial graduate institution in each great metropolitan area. In each of our emerging cities of intellect, scientific research is an essential element in building the needed institutions. Should, therefore, the granting agencies lower their standards, which now permit grants only to first-rate problems by qualified investigators, in the hope that grants to less than topnotch institutions and investigators would encourage their rise to a better status?

In my opinion, for a variety of compelling reasons, the answer is a ringing *NO!* To lower granting standards would open the door to every kind of boondoggle and soon bring the whole program down in ruins.

Instead, the problem must be attacked at its very roots. Institutions that are below standard should be motivated to initiate measures that would permit them to attain first-rate status at the graduate level. The very failure of new institutions to achieve the stature of the first 20 to 25 universities to reach maturation illustrates the depth of the problem, whose seriousness is compounded by the new needs of the technological revolution. The very purpose of the federal program is to aid in the correction of vital educational and scientific deficiencies, but clearly this aid must not be in a form to perpetuate the mediocrity that is already so comfortably flourishing.

Basically, correction must be initiated at state and local levels through the marshalling of community forces that would insist on unqualified excellence. These forces must be considerable, and developed in a variety of ways, to counter the reactionary inertia of the university which, according to the Kerr principle, can be estimated as inversely proportional to the "square" of the qualification of the faculty. The community has the right to expect minimum standards of a university it undertakes to support, and these include such things as: (a) the presence of some superior institutional leaders who understand and appreciate the qualities of creative scholarship and contribute to them; (b) an institutional plan to acquire first-rate status department by department, by suitable appointments, salary scales, teaching, counselling, and research loads and facilities; (c) an institutional plan to finance academic reconstruction, department by department, to top status; (d) a continuing institutional determination to keep scholarship clearly at the top of its list of basic activities, with all other activities incontrovertibly subordinated to this primary goal.

I have emphasized planning and action by department so that available resources can be concentrated on bringing one department to critical excellence at a time. This procedure is, of course, distasteful to those parts of the faculty awaiting their turn, unless they are especially farsighted and patient. But unless university development can be achieved within financial reason, *no* department will rise to top stature in any reasonable time. The misguided effort to make all departments all things to all people at one time has done more than anything else to prevent the rise to distinction of our middle group of universities, because limited resources are so dispersed that no department achieves this distinction.

The role of the granting agencies, it seems to me, should be a quick encouragement of community action that undertakes definitive academic planning, through promise and fulfillment of prompt federal aid and support, as local plans, procedures, appointments, finances, and administrative determination develop. Different communities in different situations must act in quite different ways, so that local initiative with quick federal response would appear the ideal.

A second policy problem of the granting agencies is the cost of research support assumed by the universities in managing grants on behalf of their faculty. This is the problem of "overhead" payments, which has been widely discussed and is ably summarized by Clark Kerr. Although it sounds dreadfully dull, the problem of overhead costs is actually one of the most critical in university development.

Most federal agencies are restricted by law from paying more than 20 per cent in overhead costs on research grants. Similar limits are not generally imposed on contracts administered by the same agencies on the grounds that contracts are given to investigators who agree in advance to do a specific job or to deliver a specific product; total indirect recovery of costs under contracts is common practice. This distinction is wholly artificial in many cases; the Atomic Energy Commission, for example, supports research by contract that the National Science Foundation would support by grant.

To the universities, the limits imposed on indirect costs of research under the grant instrument appear incomprehensible. Although it is true that the grant is designed to support basic research investigations, such research is no less demanding of administrative time, accounting, and other overhead services by the scientist's institution. Indeed, the enlightened institution will supply under "overhead" the broadest possible services in order to free the scientist for the pursuits that only he can follow. This policy increases the effectiveness of the scientist but increases the institutional overhead as well. Consequently, when such institutions undertake basic research on behalf of federal agencies, their other primary activities, that is the training and education of future generations of scientists, are penalized.

As we have seen, the increasing demands for highly trained scientists in today's scientific and technological society are already taxing the capacity of our universities to produce well-educated scientists and engineers in sufficient numbers. Moreover, the government is itself employing large numbers of our best technologically trained people, and large additional numbers are drawn into the industrial complex that supports our government programs. The Congress and the federal agencies have shown increasing recognition of their obligation to replace their withdrawals from our total intellectual reservoir; and, indeed, many agencies such as the Atomic Energy Commission, National Aeronautics and Space Administration, and the National Science Foundation have developed large programs of fellowships, scholarships, and other training mechanisms. But these same federal agencies continue to impose unrealistic restrictions on indirect costs recovery at the very institutions they are supporting, thus diminishing the universities' own resources and unnecessarily diverting the energies of scientists to activities that can more efficiently be done by others. This is a case of the federal right hand not knowing what the federal left hand is doing.

In particular, the small monies available to the nonprofit institution should be used as "seed" money, to examine and develop "long-shot" ideas at a stage where any granting agency would have difficulty in justifying support, or to develop new scientific personnel who are yet too inexperienced to merit federal aid. Indeed, in undertaking a quite new project it is often necessary for the university to finance considerable unsponsored research simply to clarify the most promising directions to be followed so that application for a grant can be adequately phrased and justified.

Operating under this costly overhead policy, an institution tends to seek ways to save administrative expense, despite the fact that supervisory expenditures are necessary to ensure sound administration of public funds. The present policy just kills the goose that lays the golden egg.

In my opinion, the Congress would ensure much greater return for the investment of public funds if it would allow at least up to 35 per cent overhead expense against direct costs under research grants, which is in the neighborhood of the actual experience of most institutions. Such a policy would: (a) release seed money from the funds now expended for unreimbursed overhead for the initiation of new and soundly conceived research projects; (b) encourage institutional support for newly emerging scientists, who are desperately needed; (c) encourage better internal administration of public funds; (d) permit more freedom of action to the scientist within his institution, with a consequent increase in his productivity; (e) reduce the reach of the federal hand in the control of our institutions.

A third policy problem relates to the duplication of large facilities. Every such duplication detracts from support of other scientific activities. Nevertheless, the very virility of science as a living activity requires active work on each scientific subject at a number of centers. For graduate training, for alternative insights, for broad national scientific strength, reasonable duplication and geographic distribution of facilities are imperative. Moreover, careful design provides variant or alternative capabilities at different facilities. Fortunately, the Office of Science and Technology has now taken an active role in studying each new major facility proposed and coordinating projects among all agencies concerned. Consequently, most such proposals are now well evaluated and justified before they are submitted to Congress.

In managing unique facilities, the operating organization is sometimes tempted to emphasize the projects of its own scientists over the needs of others. The great accelerators, ships, space probes, and reactors must be recognized as *national* facilities; good management dictates that they be operated by some agency, private or public. But wise judgment as to whose projects should be accommodated must be provided for. Since the insider tends to have an advantage, the operator of a national facility must lean over backward to ensure that the best and most productive projects receive priority, even if proposed by scientists outside his own organization. In allocating a national facility, the federal agency supporting it has a responsibility to see that good and independent judgment is exercised in determining working priorities for its projects so that they are evaluated on their merits rather than by the local interests. The management of our great accelerators has provided a wise example which can be extended to other national facilities.

A fourth major policy question relates to the form and extent of management restrictions which the federal government should impose on federally sponsored university research to safeguard the public treasury. The present policy places a heavy onus on the scientific community to recognize and accept its obligations in the use of public funds. It is a tribute to the dedication of scientists, and to the institutions of which they are a part, that so little mismanagement has occurred.

Over the course of recent years the grant instrument has

developed without explicit legal interpretations of its use but with implicit understandings between the federal agency and the grantee. It is generally recognized that research investigations, while directed toward a given objective, may (and indeed often will) lead the grantee down a path divergent from that originally conceived. Consequently, grant proposals are frequently formulated in broad terms, and the supporting agencies accord latitude to the individual scientist in the development of the specific research program. Inevitably, an unexpected result arises in the course of research, and its direction is frequently reoriented toward different end-objectives which are often better goals, providing more substantial scientific progress and discoveries than could have been anticipated before exploratory research was begun. There is no profit in following dead ends once they are recognized as such.

This process is inevitable in the pursuit of basic research, and to curtail it in the application of the grant instrument could seriously hinder the course of most fundamental investigations. The heart of the matter, it seems to me, is not whether an investigator should be permitted to change his research directions, but rather: *how far* he should be permitted to change those directions before the original considerations that led to the grant award can be regarded as no longer valid. It appears to me that this is the point of primary concern, not only for Congressional committees but for the granting agencies as well.

There is no doubt however that a few research investigators have abused the privileges accorded by the grant instruments. Their number is small, but their irresponsibility jeopardizes the whole scientific community. The question is basically one of accountability for the privilege of obtaining and using public funds. Clearly, scientists and their administrations must recognize the need for a common set of principles to be observed in the utilization of grant funds.

Scientists should take steps to see that these principles are promulgated, understood, and endorsed by research investigators throughout the nation. Adoption of such a course could well avert the need for restrictive legislative action and at the same time permit the necessary flexibility in the use of grant funds that would ensure optimum conditions for the successful pursuit of basic research. The scientific community, through the National Academy of Sciences, is now actively exploring this problem.

Certain fundamental elements are immediately obvious. For example, criteria to measure performance and scientific progress under a grant seem of the highest importance. Fundamentally, this is the problem of obtaining an evaluation of the *substance* of scientific accomplishment and the avoidance of equating progress to such measures as time or dollars devoted to the project. Research dollars should buy thought, imagination, creativity, and brilliance—not just time, which may be worthless. Surely too, it should be relatively easy to establish criteria that will demonstrate the necessity and relevance of expenditures for the purchase of equipment, special field research facilities, travel, or other specific items.

Obviously, other measures or criteria would be needed if a code of ethics is to be developed. The key objective should be to define clearly a mechanism whereby scientific responsibility is exercised by the scientists themselves and by their institutions. I would sincerely hope that such measures would reassure Congressional committees and sponsoring federal agencies that the scientific community can set its own house in order and that it is unnecessary to impose legislative constraints on the traditional freedom of research scientists. This hope arises from the certainty that overregulation would inevitably destroy the creativeness and, consequently, the present productivity of our science. Creativity is extremely sensitive to regulation and, if lost, would represent a much greater waste of public funds than would occasional cases of doubtful management. Moreover, in view of the relatively excellent history of grant management in scientific research, the cost of administering more detailed regulations could easily exceed the cost of the rare cases of mismanagement, particularly as the scientific community learns to assume more active leadership and accepts responsibility for the creation of reasonable ethical standards.[4]

A fifth policy problem hinges around the oft-heard comment or criticism of "big science" or "big research." The implication is that any effort in scientific research beyond some undefined size is somehow bad for science and for the universities, and that government is acting improperly in funding such big science.

The fact is that different kinds of scientific activity require quite different scales of effort. Astronomy, nuclear physics, oceanography, and space science all require large basic expenditures and some measure of teamwork to provide experimental results. Moreover, such sciences as certain aspects of geophysics require coordinated, synchronized, and well-planned observations of many scientists acting together. Nevertheless, even in the biggest scientific projects or laboratories, one finds the major scientific activities composed of small elements being studied more or less freely and independently by small, enthusiastic groups of scientists, just as in smaller institutions. Certainly the IGY was big science, comprising thousands of smaller coordinated researches, yet it was an unqualified success as big science, for it introduced man to his home—as a planet!

Clark Kerr points out that big science (and technology) is simply a natural response to the scientific revolution of

4. See report of the National Academy of Sciences, "Federal Support of Basic Research in Institutions of Higher Learning," Washington, 1964.

our time. Science has become big where it is necessary to solve big problems, and the results have repaid us handsomely. To rail against big science is akin to a demagogic attack on sin, which is not very clearly defined and probably not there at all when you try to find it. Certainly, unwarranted expenditure in science, whether big or little, is intolerable and should be searched out and eliminated. The universities managing the elements of big science are also among our best institutions, and I would trust their judgment in its development.

A sixth policy problem relates to the payment of graduate students under research grants or contracts. For many agencies this is forbidden by law. We have already seen that graduate education and research are inextricably entwined. In many cases the best way to do research and to get the results is through a graduate student who, with proper counselling and guidance, will win his scientific spurs and his degree at the same time. This is a problem that should be distinguished clearly from the issue of "federal aid to education at the undergraduate and secondary level." It is a problem that clearly involves our posture for national defense. In this new age, when every doctoral graduate is literally worth his weight in gold to the community,[5] no law or regulation could be better designed to waste money or to deprive the community of the chance to get both research and graduates at one price. Research grants should *encourage,* not discourage, the affiliation of graduate students and postdoctoral "interns" with *all* programs of fundamental research.

5. Many doctoral candidates weigh 200 pounds (avoir.). This is about the same as 2,916 troy ounces. With gold pegged at $35 per troy ounce, the candidate's weight would be valued at $102,600. With his doctoral degree, he should earn about $250,000 more in a lifetime than the bachelor in similar technical fields. He probably would contribute more than $1 million to the community in leadership. Many do better!

Likewise, our great national laboratories should be encouraged, not discouraged, to enter more actively into the training of graduate students in cooperation with universities or other organizations with suitable academic standards. Here we have institutions of the finest quality, which at small cost could reproduce their kind many times over, to the great enhancement of our national strength.

Finally I would comment on the diversity of federal agencies administering grants and contracts for scientific research. In my opinion, experience has shown the present system best fitted to the American scene. It is outstandingly successful. Diversity is the essence of our American system. We can only be impressed with the advantages that have become apparent in the support and administration of research by each of the several departments and agencies of the government, each in its own area of activity:

1. Each department and agency is kept close to the advances of American science most intimately affecting it and is progressively influenced by that participation in scientific activity. Such participation counteracts the inevitable tendency toward obsolescence in government so evident where its agencies are insulated from science. The influence of science throughout our government is now very strong and productive of youthful, virile attitudes that are in striking contrast to the dead bureaucracies so often seen abroad.

2. Since each department and agency functions primarily in its own areas of interest, its leadership is automatically equipped with an inherent, "built-in" comprehension of the research it must administer. This ensures efficient administration and effective evaluation of research results.

3. Because the form of administration and regulation of research grants and contracts varies somewhat from agency to agency, American science has had the opportunity to experiment with different types of administration. As a re-

sult, steady improvement has occurred, and bureaucratic authoritarianism in one agency can make little headway in competition with contrasting methods in other agencies.

4. Diversity in the administration of research grants tends to reduce the hazards of scientific orthodoxy because the administrative point of view varies among agencies, and the broader aspects of science have greater opportunity to be recognized for their worth. This gives us great national scientific strength.

Professor A. Hunter Dupree[6] has traced the rise of pluralism in the federal sponsorship of scientific research, a history that extends back for more than a century. He points out that,

> The scientific community has a fundamental stake in pluralism and should be expected to encourage it in its relations with the larger community. The scientific community should [protect] the institutions through which pluralism is manifested, so that variety and flexibility will be characteristics of the scientific community itself.

Dupree goes on to analyze the forces impelling toward a central federal scientific organization throughout our history, and points out the inevitability of the Office of Science and Technology and the Federal Council of Science and Technology, to provide needed coordination as science became large. Under pluralism, we must recognize that the danger of unnecessary or unplanned duplication is enhanced, and serious gaps in support planning are inevitable, since no one operating agency has the responsibility for all of science.

The response of our government in creating these new coordinative and advisory agencies under a Presidential

6. A. Hunter Dupree, *Central Scientific Organisation in the United States Government, Vol. 1, Minerva,* London, Summer 1963.

Science Advisor and the President's Science Advisory Committee is a typical American response to the desire to retain the advantages of pluralism and at the same time establish a control mechanism to keep it from getting out of hand.

Some years ago, because of the glaring lack of support in the important area of geophysics, I advocated the creation of a department or agency for the environmental sciences. Subsequent events have erased this need. Among these events were, for example, the massive support for and the outstanding performance of the International Geophysical Year, the impact of the Sputniks, the recent support for oceanography and the Mohole, and the creation of the Space Agency.

I have dealt at some length with the impact of federal aid on the scientific research done at the universities because of the enormous effects of that aid on the evolution of the universities. To some people, any change is bad. But the dramatic birth of our scientific culture with its new economy requires an equally dramatic response by education. The Federal Government has provided the means. Some twenty or so leading universities have provided a realistic response. Neither science nor government need apologize for the expenditures that have been made. The impact of the results of these expenditures on our economy, and indeed on the economy of the whole world, has been enormous, for these funds invested in the sciences have been returned many times over. In the field of health, we need only read the report of the Lasker Foundation[7] to see the immense improvement in the last ten years. Our national defense is second to none. No expenditures ever made by the Congress have suffered so few criticisms. Although we must make every effort to avoid and minimize errors, we must not permit criticism of the few inevitable errors that do occur to reflect

7. "Does Medical Research Pay Off? in Lives? in Dollars?" National Health Education Committee, Inc., 1963.

adversely upon what has been an enormously successful program in our national development. Rather we should strive to perfect our procedures. We have been fortunate in the leadership that could foresee the place of scientific research as a new resource which would revolutionize our economy.

In this tremendous effort we are witnessing the response of a great nation to an almost overpowering social change. That mistakes will be made is inevitable, and these should be viewed and evaluated, not in a narrow context, but in the perspective of the larger objective of social readjustment. The greatest and most costly errors today are those of omission, not commission. In the dramatic development of our new economy, we cannot forget that less than half our population is adequately served by distinguished research institutions and graduate universities. As Clark Kerr concluded: "The major test of the modern American university is how wisely and how quickly it adjusts to the important new possibilities. The great universities of the future will be those which have adjusted rapidly and effectively."[8]

Having embarked as a nation on the course of federal support of science and technology, out of which an economy of plenty has been derived, the future stability of that economy rests unambiguously on the wisdom of the policies that will govern the future of federal support. The recent (1964) reductions in federal aid to basic scientific research are a danger signal that there is some lack of comprehension of the relation of science to our new economy. With a reduced science, and eventual and consequent reduction in imaginative innovation, unemployment and relief rolls could bring our economy of plenty tumbling down around our ears. The major problem today in terms of federal support of scientific research is how to extend a virile science throughout the length and breadth of our nation.

8. *The Uses of the University.*

# 4. Science and Philosophy

In the witty and penetrating discussion that constituted his Rede Lecture at Cambridge University in 1959, C. P. Snow directed attention to a major social problem. A well-known British novelist with a background of work in the sciences while a fellow at Cambridge, Snow chose the title "The Two Cultures and the Scientific Revolution." Let me recall Snow's thesis:

> At one pole the scientific culture is really a culture, not only in an intellectual, but also in an anthropological sense. That is, its members need not, and of course often do not, always understand each other . . . but there are common attitudes, common standards and patterns of behaviour, common approaches and assumptions. This goes surprisingly wide and deep. It cuts across other mental patterns such as those of religion, or politics, or class. . . .
>
> At the other pole the spread of attitudes is wider . . . But I believe the pole of total incomprehension of science radiates its influence on all the rest. That total incomprehension gives much more pervasively than we realize, living in it, an unscientific flavor to the whole "traditional" cultures . . . if the scientists have the future in their bones, then the traditional culture responds by wishing the future

did not exist. It is the traditional culture, to an extent remarkably little diminished by the emergence of the scientific one, which manages the western world.

This polarization is a sheer loss to us all. To us as a people and to our society. It is at the same time a *practical* and *intellectual* and creative loss, and it is false to imagine that these three considerations are separable.

Snow is undoubtedly correct in defining a scientific culture. This is an age in which science with its own highly defined philosophy dominates our culture. The basic tenets of this philosophy began to emerge during the sixteenth and seventeenth centuries in the brilliant works of Copernicus, Brahe, Kepler, Galileo, Newton, Boyle, Hooke, and their many contemporaries and successors. Of it, John Donne[1] exclaimed, "And new philosophy calls all in doubt." The "new philosophy" endured and profoundly lifted man's general culture and material welfare within the relatively brief span of four centuries.

The scientific philosophy is not teleological in the sense that all subsequent thought must be rationalized to satisfy some initial or intuitive premise. Revelation or supernatural intervention has no place in science. The scientific philosophy is a *method* of problem solving in which the final end cannot be foreseen at all, nor indeed is any final "cause" (in the Aristotelian sense) assumed or admitted. It is the very antithesis of the method of Plato[2] in which we observe a brilliant intellect attempting to fathom the plan of the creator of the universe. He believed that man had no surer guide for this task than to follow the indications of his particular personal sense of the harmonious, the beautiful, and the desirable and that "on these subjects we should be satisfied with the probable, and seek no further."

1. *An Anatomy of the World.*
2. *Plato's Cosmology, The Timaeus of Plato,* tr. and ed., M. F. Cornford, Harcourt, Brace, New York, 1937.

The scientific method is a means whereby a stable, generally agreed upon, ever enlarging body of knowledge can be generated in a form that can be independently checked and reconstructed by anyone familiar with the method. Knowledge derived by the scientific method is not discarded as a result of subsequent discovery but is absorbed into larger and more encompassing conceptions as they are formed. Although scientists in different fields may speak quite particular and specialized languages, they are united in an unswerving adherence to the scientific philosophy.

What are some of the principal elements of this scientific philosophy?

1. The ideal of exact quantitative descriptions and laws whenever possible.

2. The principle of simplicity; that is, the simplest physical model that connects a series of facts is most nearly the truth, and therefore represents the greatest beauty.

3. The principle of generality; a conception is useful in direct proportion to the breadth of factual knowledge that it embraces.[3]

3. In epistemology (the theory of knowledge) there has been widespread discussion concerning the distinction between sensory experience and conceptual description. Einstein believed that the world of sense impressions is comprehensible in the following sense: one fashions general concepts, and relations between them and imposes relationships between concepts and sense impressions, and thereby achieves an order among sense impressions.

According to Kant, the positing of a real external world would be meaningless without such comprehensibility. The mode of assigning concepts to sense impressions is decided only by success in establishing an order of sense impressions. The rules of correlation can never be definitive, but can claim validity only for a specified region of applicability. (See *Albert Einstein, Philosopher-Scientist*, ed. Paul Arthur Schlipp, Library of Living Philosophers, Evanston, 1950, esp. "Einstein's Conception of Science," pp. 385–408, F. S. C. Northrop; also Victor F. Lenzen, *The Nature of Physical Theory; A Study in Theory of Knowledge*, Wiley, New York, 1931.)

4. The principle of hypothesis to guide the selection of experiment (in order to circumvent the manifestly impossible task of performing each of all conceivable experiments), whereby a model embodying the known facts is assumed and then tested for its validity through the most rigorous experiments that can be conceived. The hypothesis is modified, confirmed, abandoned, or absorbed in a larger generality, in the light of new experimental data, and only accepted as law within circumscribed limitations when no exception within those limitations can be discovered.

5. The absolute dependence on experiment; no conception or hypothesis is worthy of consideration unless it can be questioned critically by experiment that will yield confirmed results by independent observers.

6. The principle of control, whereby the rigor of an experiment arises from its power to separate clearly and to define independently each contributing parameter.

7. The principle of synthesis, whereby a wide range of simple interactions is first explored and defined, and broader generalities then induced from the related behavior of the assemblage.

8. The ideal of absolute honesty whereby the order of each perturbing effect is ascertained, no perturbation of similar magnitude is ignored, and the conclusion is circumscribed by the limitations of the experimental data.

The scientific philosophy operates on the basic premise that the entire universe as well as its component and elemental parts are governed by an invariant and ascertainable order, equally applicable everywhere, whose behavior is rigorously describable and predictable within specifiable probability distributions. Hence, all events are explicable by Aristotle's "efficient causes." Over the past three centuries, these elements of the scientific philosophy have been tested rather thoroughly.

The first principle, that of quantitative description, not only permits prediction of the future behavior of a system, but it immediately distinguishes important from unimportant factors in any situation. One of the great criticisms leveled at the social sciences is their all too frequent inability to distinguish first-order from second- and third-order effects within a society exposed to a variety of forces. In any system it is quite useless to focus attention on minor phenomena when they are overwhelmed by a major phenomenon whose effect is often not defined or comprehended or is sometimes deliberately ignored.

Insistence on quantitative description separates the sheep from the goats among ideas, with a conclusion not infrequently amazing to everyone concerned. For any assertion, one has the right to ask, "What are the numbers?" That is, does the quantitative description of the assertion show that it is of first-order importance or of real relevance with respect to other independent and major factors in the situation?

The principles of simplicity and generality of the scientific philosophy are particularly vital. The Ptolemaic description of the heavens, conceived around a geocentric model, satisfied the scientific requirement for a quantitative definition of their behavior. The Copernican table could hardly improve on the Ptolemaic system, which was easier for people to believe because it satisfied the evidence of their own eyes. But as Galileo showed, the conceptions of Copernicus, with Kepler's modifications, were both simpler and more direct, and therefore much more general. Thus, in the scientific sense they had greater beauty for *man could not perceive the three basic laws of motion subsequently postulated by Newton until Copernicus had conceived the heliocentric system with its natural simplicity and generality.* This story is repeated again and again through the history of science. The mere mathematical description around "any old" model is not enough. The model must meet the criteria of simplicity and generality if it is to become an instrument for the further advancement of human understanding.

Reliance on controlled observation and experiment is central to the scientific philosophy. The design of the experiment determines its rigor in avoiding undesired perturbations that otherwise obscure the results. In the early days of modern science, only rather elementary experiments could meet the requisite standards. But today, with the great advances in applied mathematics, probability theory, and computer practice, genuine experiments that are fully qualified in the scientific sense can be and are being designed to inquire into very complex situations.

For example, the scientific philosophy can now be broadly and intelligently applied to the social sciences. With the use of computers, it is becoming increasingly possible to construct complex models that simulate a variety of social problems that scientists would otherwise have no practical way of studying.

At present no one can say how deeply the scientific philosophy can invade the realm of human affairs or rationalize the primitive and ad hoc philosophies under which civilization has risen. Certainly the end is not yet in sight, and new scientific tools and experiments constantly extend its horizons. Each excursion yields reproducible knowledge on which men can agree. Indeed, the very extent of this invasion is now contributing to the profound social changes so apparent all around us.

A common misconception is that science consists of the rather routine observation of numbers or facts and their mere classification. Such procedures are involved, of course, but in a relatively subordinate way. In its largest sense, science involves a growing curiosity about some unexplained behavior and the creation of hypotheses relating that behavior to known events—hypotheses that intelligently direct further experimental inquiry. Science demands ingenuity in the design of experiments or the perfection of observation that can isolate and distinguish each measured quantity from superimposed perturbations; it involves cutting through complexity to reveal the simple elements from which complexity is synthesized; it involves the successive reassociation of facts in a multitude of ways that enable the assessment of the relative strength of each combination.

Science is creative beauty in the highest sense. It provides a systematic and reliable criterion of universal applicability in Plato's search for "the harmonious, the beautiful, and the desirable." The modern painting with its confusion of forms and color, through which the underlying theme can at first glance scarcely be perceived, is reminiscent of the frantic search in the mind of the scientist for the central connecting order in a profusion of vaguely related ideas and facts. The search by the modern composer for the ultimate harmony of dissonant chords reflects the excitement of the scientist in his association and reassociation of ideas during the intuitive hunt for the coherent generality that represents a higher order of human thought and comprehension.

Truly, science and art are related intimately in the growth of fuller understanding. Who knows when some artistic work may provide the clue to some higher formulation of human comprehension? For the symbolism that can best grasp and represent ideas goes far beyond the written word and constantly takes on new and improved forms. Herein lies the glory of mathematics.

In closing this abbreviated discussion of the scientific philosophy, I would reiterate its utter dependence on experiment for its development. No assertion or assumption, no logical derivations can be accepted as truth unless they are capable of experimental verification within defined limits. Scientific experience has demonstrated repeatedly that no speculation or assumption, however reasonable, can be accepted as truth without such verification and that ideas must be formulated in a manner that permits their clear and unambiguous testing. The unanticipated perturbations arising from such critical tests give clues to new and unperceived forces whose recognition enlarges the scope of human thought.

In turning to the broader problems of the philosophy that guides a society, in our endeavor to find the appropriate emphasis on the body of scientific thought, I trust that the scholars in the field will forgive both my brevity and the elementary nature of my discussion. Yet the almost complete lack of familiarity with scientific thought in the liberal education and thinking of our time suggests that we should review the problem in a rather simple and direct way.

Now, obviously, at any instant of time, a society must have a set of generally agreed upon rules and conceptions with which to govern relationships among men. These rules and conceptions are derived from its philosophy—a body of rather fundamental and coherent thought—out of which reasonable and hopefully rational judgments can be rendered and decisions can be made in any given situation and in a reasonably consistent way. Members of the society must be acquainted with this philosophy so that each can reasonably anticipate the behavior of the others. The problem of any society is: "What philosophy shall govern the decision-making process either of the individual or of the body politic?"

We recall clearly the rise of humanistic thought in the Renaissance as a revolt against the straitlaced philosophy of piety and other-worldliness emphasized by medieval scholasticism. Humanism was a philosophy that dignified the individual. Humanism asserted that, in the decisions of society, the rights of the individual to beauty, compassion, and justice must be respected. The revival of the humanist tradition, after the dull medieval interlude, emphasized the human goals that made life not merely bearable but really worth living. The humanist philosophy is a proud one, lifting the human spirit from slavery.

But the decisions made within the framework of a humanist philosophy must also be rational; that is, they must not obviously violate the laws that govern the universe or the accepted relationships among men. Thus the humanist tradition is not a philosophy of license but one of agreed upon discipline leading to the optimum freedom and opportunity for each individual within the collective framework of that society.

In absence of any consistent body of scientific thought, a philosophy can only be based on synthetic criteria, derived from the more obvious manifestations of man's experience —his uncritical and intuitive observations. A primitive philosophy by definition is almost wholly a body of thought and belief derived a priori—a world conceived largely in terms of mutually disconnected phenomena.[4] As a body of scientific thought develops from an independent but completely rational scientific philosophy, it constantly encroaches on the more primitive general philosophy of the society through the substitution of new, reliable, and completely rational ideas. Science thus provides a dependable means for determining action, where before only ad hoc assumptions were accessible.

But at no time is the rational content of science sufficient in itself to provide a complete philosophy that can govern all the decisions that society must make. Let me underline this point unambiguously! The body of human thought generated by the scientific philosophy is generally speaking un-

4. I use a priori here in the same sense as Kant, referring to presumptive knowledge that is given, grasped, or conceived, independent of explicit analysis of sense impressions.

assailable, but its content can never be sufficient to govern human relations in their totality. Thus there must always remain scientifically unexplored, or unexplorable, areas of experience for which ad hoc, but generally agreed, estimates are required in governing human relations. Quite outside the scientific philosophy at any time there remain great depths of human response for which we must turn to literature, art, music, and theology.

Nevertheless, the completely rational aspects of human experience, represented by the ever growing body of agreed scientific thought, must be continually incorporated into a society's total philosophy, replacing the ad hoc assumptions previously made. This process of articulation of rational scientific thought into the total philosophy should not displace humanism; it should give humanism a broader base from which to rise to greater heights.

It is on this point that I believe Snow to be on thoroughly solid ground. Our society is faced with a dilemma—the polarization of our philosophy. This is a wholly artificial polarization separating into two parts what should be a coherent whole. On one hand there exists a now large and completely coherent body of scientific knowledge which can be made to yield the predictable consequences of rational decisions. On the other hand, there remain perpetuated philosophic areas *covering precisely the same ground* that were derived in earlier history from a priori assumptions which were necessary at a time when no better guides were accessible. One is led to quite different conclusions and decisions, dependent upon which philosophy is chosen.

The conflict between the rational and the precedent a priori philosophies has a long history, with sometimes disastrous consequences to society. The recantation of Galileo, the history of witchcraft from the imprisonment of Kepler's mother to the Massachusetts Colony, the deliberate book burnings by the medieval world to erase knowledge of chemistry and metallurgy, the long conflict with Darwin's evolutionary theories and the pursuant Scopes trial in Tennessee are but a few of the more obvious examples.

Of course, with time and the diversity of world cultures, such philosophic errors are inevitably and eventually erased. But that they have occurred has slowed the rise of culture and perpetuated ignorance, poverty, and brutality far beyond their time. Today the conflict proceeds on more subtle grounds, yet its consequences are equally damaging to our culture. The reasonable course would be the continual articulation of these two poles of philosophy into a single coherent whole, employing the maximum power of scientific thought as it is continually enlarged.

Who is responsible for the present dichotomy—the humanist? the educator in liberal arts? They have just taken tail and lit out, leaving no one to bear the proud torch of humanism in form suitable to the best interests of today's society.

It is this dilemma that leads Schroedinger in his Dublin lectures, "Science as a Constituent of Humanism," to plead:

> The majority of educated persons are not interested in science, and are not aware that scientific knowledge forms part of the idealistic background of human life. Many believe—in their complete ignorance of what science really is—that it has mainly the ancillary task of inventing new machinery, or helping to invent it, for improving our conditions of life. They are prepared to leave this task to the specialists, as they leave the repairing of their pipes to the plumber . . . the fifty years that have just gone by . . . have seen a development of science in general, and of physics in particular, unsurpassed in transforming our Western outlook on what has often been called the Human Situation. I have little doubt that it will take another fifty years or so before the educated section of the general public will have become aware of this change . . . this

process of assimilation is not automatic. *We have to labour for it.*[5]

The duality of great areas of human thought in our time, one wholly rational, the other haphazard and often irrational, has led to the artificial polarization of today's humanism and science. This is tragic in the intellectual sense, but it is also downright dangerous to a society which now leans so completely on science for its social and economic development. Such dichotomy of philosophy is all the more intolerable because the "body politic" must make a large variety of decisions which ought to depend almost wholly upon the assessment of the facts derived from the scientific philosophy. Indeed such decisions are becoming the major decisions of our time. Whether to put a nuclear reactor near a city is not an emotional question. It is a question to be decided by weighing the advantages against the risks in the light of available safeguards and possible alternatives.[6]

Should we open up a canal by means of nuclear explosions? What are the really relevant factors and risks in a nuclear test ban? What safeguards are needed in the use of pesticides? How far should our food and drug laws go? These are questions to which objective answers can be found if the public has some degree of awareness of the quantitative aspects of the scientific philosophy—a public that insists on weighing the absolute "pluses and minuses." In absence of that familiarity, decisions are governed by emotion, demagoguery, and chance.

5. Erwin Schroedinger, *Science and Humanism,* Cambridge University Press, 1951, pp. 9–11.

6. Too often the decision is based on an unreasoned fear of "nuclear hazards" without taking into account the hazards of conventional (but accepted) atmospheric pollution from the burning of fossil fuels. Since all human activity involves risk, the problem of minimizing risk and maximizing human benefit is primarily analytical.

For example, probability theory plays a major role in the life of every individual. In absence of any knowledge of that theory, he is forced to learn by crude trial and error. He accepts the probability of death by automobile out of his experience that the risk is apparently compensated by the effectiveness, convenience, and pleasure it adds to his living. But at the same time he rejects unfamiliar activities involving much less risk and potentially adding much more social profit, since he is not educated to make an objective assessment, nor does he know how such an assessment should influence his attitudes. Only some reasonable familiarity with the scientific basis for our new environment can remove the emotional stress that too often leads to improper decisions. In particular, in light of today's knowledge, the most "human" course of action may not be at all obvious to the public without public attitudes that require, and will accept, the analyses of the alternatives at the limit of our competence.[7]

But fundamental questions are now being posed to our new science-dependent culture. What is the real meaning

7. Some humanists may very well attack at this point by observing that, even after skillful analysis, scientists will not agree among themselves on many public issues. Of course they don't! As noted earlier, science will never embrace all knowledge, so areas for debate are always open. The point is that debate should not rage on grounds where rational knowledge is available, and would not, if our philosophy and education provided for prompt recognition of that growing body of knowledge. Nor do I believe the point should be discredited because individual scientists, like all other humans, evince areas of ignorance, obtuseness, or prejudice. (Cf. Robert M. Hutchins, Scott Buchanan, Donald N. Michael, Chalmers Sherwin, James Real, and Lynn White, Jr., on "Science, Scientists, and Politics," an occasional paper in *The Role of Science and Technology in the Free Society,* published by the Center for the Study of Democratic Institutions, 1963.) To live in this scientific age, we should grasp all the knowledge at hand, and employ it with the greatest possible effectiveness.

of our population explosion in terms of future human welfare? What must be the new proportions in our "mix" of educational capabilities to keep virile our new economy of plenty? What are the elements in a national policy that will assure adequate support of scientific research in order that we may realize both the potential growth of our new economy and achievement of our intellectual goals? Are we convinced by the examination of such problems that new forms of action are imperative or that we must pay more taxes for specific activities in order to achieve higher economic or social levels? Is foreign aid really necessary to the underdeveloped world, and does its form make any sense in terms of the new scientific culture? As poverty is lifted, what new human goals should be set in response to our new environmental situation? Now these are questions whose answers are viewed quite differently by a culture dominated by the humanism of the Renaissance and by a culture incorporating the capabilities of scientific thought today.

There is a certain irony in the fact that objectives which men of high moral purpose have labored through millennia to achieve are made possible overnight by the efforts of the scientists responding impersonally to the forces of society. The desire for peace, for example, has found universal expression in every age, and it seems in the main to be secured in our own time by weapons so awesome that no nation dares take them up against another.

The theologians have sought to fortify men's spirits and lighten their temporal burdens of poverty, illness, and toil by holding forth the vision of a better world to come. But science and technology have amply demonstrated that it is possible to eliminate poverty and disease and to create a good life that rivals the dreams of the prophets. Perhaps the task that remains is to teach us how to impart the capacity for creating the good life to the vast underprivileged areas of the world.

92

Science has had to be flexible in responding to what is moral in one generation and immoral in another. In an earlier age, when populations were regularly decimated by wars, pestilence, and famine, human fertility was encouraged as a high moral objective. Now that science has eliminated so many of the ancient threats to survival, the human race is expanding at a rate that threatens to extinguish itself with its own numbers. So now we turn to science to find ways to curb the population explosion that are not only feasible but morally acceptable to men of widely varying faiths.

But beyond this, as science provides a new rationalism to our social philosophy, our ancient attitudes must continually adjust to this new rationalism, so that society will continue to benefit from the gains that have already been won and at the same time move on to new achievement. What was moral before, often becomes immoral in an advancing society.

In the early stages of civilization, when change was gradual, culture could be represented to a sufficient degree of approximation by a series of cultural steps, each stable for a millennium or at least a few centuries. In such a framework Aristotelian "final causes" could be established, around which stable social dogmas could be erected, since they would be socially dependable for many generations. But the speed of social change today inevitably forces the humanist and the philosopher for the first time to face up to the calculus of motion. When the mathematical "derivative" of social development was imperceptible, human affairs could be viewed within a framework of successive static steps. But when that derivative is so large as to revolutionize human affairs in each generation, then the old static guidelines become useless. Philosophy has thus been pushed into a relativistic framework, in which morals and dogmas, beauty and harmony can only be measured significantly from a rapidly and radically changing base. The humanist is now inevitably enmeshed in the mathematical vortex of this new social relativity. In absence of a coherent and all-embracing body of liberal education he has neither the means, nor can he find the will, to comprehend or to rationalize society in this utterly new environment.

Consequently, we often see day-to-day events interpreted as isolated and unrelated power plays to benefit individual opportunists, when in the broader context they form vital pieces in the growing structure of our social strength. In a rapidly moving society, the constant experimentation that is the core of a free system of diversity requires a highly rational system of judging and selecting the alternatives. Competition of the new, as contrasted to older, methods makes essential an informed judgment to avoid improper decisions that will delay social development—retrograde decisions which, out of ignorance or nostalgia for ancient values, have evaluated the debate quite incorrectly.

In our now relativistic society, not only do philosophical values change in magnitude but, even more uncomfortably, they all too frequently turn up with the wrong algebraic sign. Thus the ancient morality, which was soundly founded in its time, now impels a population explosion whose exponential spells disaster.

In absence of a really broad and well-organized liberal education encompassing the basic ideas and methods of science, society is left without the means of intelligently nourishing the science that supports its present culture or of utilizing the power of that science most effectively. Out of ignorance of the character of this vastly expanded intellectual resource, decisions fatal to society's welfare are likely to be made quite innocently—decisions relating to education, science policy, or to novel applications. The point is that liberal education in the traditional sense no longer fits man for the decision-making process that his present environment requires.

Gerard Piel, publisher of *Scientific American,* in his lecture "The Acceleration of History"[8] extends graphically the early studies of Henry Adams showing the rate of change in a wide variety of human affairs. Plotting this increasingly rapid rise against a reference baseline that relates today's rate with the levels of activity of the distant past, he sees today's changes rising exponentially toward infinity. He concludes:

> The old regime of scarcity is at an end; the time has come to repeal the iron law that says that one man's well-being can be increased only at the expense of his brothers. In its place we must frame our values and institutions to respond to the new dispensation of abundance.

To maintain a viable society in an exponentially rising tide of human affairs, it is clear that the baseline of thought against which action is measured *must also change exponentially*. From this ever new perspective, modern man will view the changes quite differently than would some primitive people. But the needed change in base can be acquired only from a liberal education that is very directly coupled to the sources of change.

The problem of liberal education is to maintain the baseline of thought in relation to the change in human affairs so that the differential between them does not become unmanageable. A relativistic view of human problems involves the necessary transformation in the coordinates of the frame of reference from which change is measured. Piel is right in that no exponential can rise indefinitely—the solution is to advance the framework of thought at an equal pace so that human perspective remains reasonably stable.

8. At Phillips Academy, pp. 22–23; April 1963.

Fundamentally, the problem of creating a stable world society on a rapidly changing baseline should rest with men of letters through their influence on society and especially on political leadership. Here the university, more precisely the individual members of its faculty, should accept responsibility for facing the problems intelligently and endeavoring to formulate new principles from which useful rules of human behavior can be derived. In this connection, one recalls the comment of Wolf[9] on the emergence of the new science in the sixteenth century:

> The Universities might have been expected to lead, or at least to share, in this movement for intellectual emancipation. But they did nothing of the kind. Philosophy was only tolerated as the handmaid of theology. . . .
>
> It was indeed highly significant of the times that the vast majority of the pioneers of modern thought were either entirely detached from the universities, or loosely associated with them. New organizations . . . were necessary to foster the new spirit, and enable it to express itself.

From a rekindled humanism, science itself might receive its greatest inspiration. In Samuel Johnson's *Rasselas*[10] the prince comes upon one of his master craftsmen one day as he is completing the building of a sailing chariot. Responding to the prince's approving notice, the artist declares:

> Sir, I have been long of opinion that, instead of the tardy conveyance of ships and chariots, man might use the swifter migration of wings; that the fields of air are open to knowledge, and that only ignorance and idleness need crawl upon the ground.

9. A. Wolf, *A History of Science, Technology and Philosophy in the 16th and 17th Centuries,* The Science Library, Harper, New York, 1950 (reprinted 1959).

10. *Rasselas, Prince of Abyssinia,* Chap. 6, ed. Grant McColley; Packard, Chicago, 1940.

And then the artist presciently adds:

> You will be necessarily upborne by the air, *if you can re-new any impulse upon it faster than the air can recede from the pressure.* [Emphasis added.]

An informed humanism gives mankind a perceptive imagination. From a rejuvenated humanism, linked with the natural sciences, one would expect a new and potent kind of social science to develop. Here the signs are encouraging, with a number of men of real stature already leading the way in the first tentative steps toward a sound science of society. True, one still finds in the woods of social science a few fuzzy creatures, but significant work is being done that meets the criteria of real science. The rise of the behavioral sciences is marked by the debate today within the National Academy of Sciences whether they should be recognized as bona-fide sciences by electing to the Academy their more perceptive practitioners. Already scientific psychology and anthropology are so recognized, and I would anticipate that very soon econometrics and others of the more exact social sciences will be formally joined to the ranks of true science. The task ahead is the erection of a sound body of social science constructed within the framework of the scientific philosophy.

In a number of other directions, though not very spectacularly, there are distinct movements toward the enlargement of a scientific culture to match today's social change. Through the Science Advisor to the President and the Office of Science and Technology (within the Executive Office of the President), analytical judgment based on scientific methods is gradually becoming an input into the governmental process. An official at the level of assistant secretary is now responsible for science and technology in each of the major government departments and agencies. Science attachés advise our missions abroad, although this innovation still is embryonic. Significant studies of the impact of science on specific social objectives are being made by the National Science Foundation, the National Institutes of Health, and the Census Bureau.

But the input of the scientific method into that large area of social development represented by the Federal Government is still discouragingly small. Major programs such as foreign aid are undertaken without the benefit of operational research, and scientific leaders in government are denied funds to initiate scientific study of their social responsibilities —studies which on one hand would vastly improve the effectiveness of government expenditures and on the other could save lives and billions of dollars in tax money. Government is hampered by lack of public confidence in the social sciences because of unbrilliant and poorly disciplined thinking in these areas in the past. The small but effective leadership in the genuine sciences of society should be hailed for its courage.

Perhaps surprisingly, corporate and community leaders are making a very direct attempt to understand the nature of our changing social situation. Through their leadership at least a dozen states have a variety of committees and commissions looking into various aspects of the new culture. I am aware of such activities in California, Florida, Minnesota, Michigan, New York, Ohio, Oregon, Pennsylvania, and Texas, and there are doubtless others.

In Detroit, for example, the officials of Wayne State University and of the municipal government are joining to redevelop the whole central portion of the city into an intellectual and scientific research center. Here the endeavor is to articulate the intellectual goals of the university with the economic and cultural goals of the metropolis. Great areas of slums are being replaced by enlarged university facilities, pure and applied research laboratories, and a variety of innovations designed to tie together the central goals of the metropolis.

My own institution, the Graduate Research Center of the Southwest with its academic arm, the Southwest Center for Advanced Studies, was founded by men who understand that high scholarship is the wellspring from which the new culture, with its social and economic advantages, must flow. The Center looks ahead to exercising a larger influence on a region containing some twenty budding metropolitan centers.

From a variety of experimental approaches, natural selection is certain to produce new solutions which can be emulated as they emerge. But our society might well stimulate a diversity of such experiments. Leaders in physics, mathematics, chemistry, and biology have been making an exhaustive review of our elementary and secondary teaching of science. A whole reorientation of the presentation of the subject matter and methods of science is taking place. The new courses are being tried out in a type of "operational research" to ascertain how the student really learns best. The old method of presentation, based on the chronology of the discovery of knowledge, is being abandoned. Instead, the new teaching emphasizes what is most important in the subject and how the whole body of science is articulated.

It has been found that problem solving, utilizing a variety of ideas from science and mathematics, can be successfully introduced at the very beginning of the educational process. The immense body of scientific knowledge is being compressed and presented by expert teachers through visual–aural aids in illustrated lectures which, in their preparation, receive the most careful attention; these lectures can be made available on a nationwide basis. The local teacher now has time to assume the added role of counsellor and to assist the students to proceed at a rate consonant with the ability of each. One can hope that similar efforts can be encouraged at the university level, with emphasis on idea content rather than mere form of presentation.

In music and in the arts a few pioneers are beginning to reflect in their work the tremendous impact of the scientific revolution upon our times. This year the compositions of the University of Michigan's Ross Finney will be introduced by the orchestras of Paris and Philadelphia. One must admire the efforts of such men as they look to the future for the inspiration for their art.

On the whole, and with a few notable exceptions, the scientists too are aware of the culture of the future, and they are responsive to the admonitions of men like Father Hesburgh of Notre Dame:[11]

> I grant you that many humanists, jurists, philosophers and theologians are illiterate in the vast and growing area of modern science and technology. Their illiteracy in your area is no argument for your illiteracy in their field. I am not excusing them; I am only trying to make the centrality of your position in the world today more fruitful, more meaningful and more significant in its total effect.

We are finding the Alfred North Whiteheads[12] and the Eric Ashbys[13] but all too few. Bronowski and Mazlish[14] re-

11. From "Thoughts of Our Times," three addresses by Rev. Theodore M. Hesburgh, C.S.C., President, University of Notre Dame.

12. Alfred North Whitehead, *Adventures of Ideas,* Macmillan, New York, 1933; also his *Reflections on Man and Nature,* with Prologue by Ruth Nanda Anshen, Harper, New York, 1961.

13. Eric Ashby, *The Technology of the Academics,* St. Martin's Press, New York, 1959; *Community of University,* Cambridge University Press, 1958.

14. J. Bronowski and B. Mazlish, *The Western Intellectual Tradition,* Harper, New York, 1960. See also J. Bronowski, *The Common Sense of Science,* Harvard University Press, 1955; and *Science and Human Values,* Harper, New York, 1959.

analyze the philosophy underlying *The Western Intellectual Tradition*. In *The Roots of Scientific Thought,* Wiener and Noland[15] attempt a cultural perspective in light of the technological explosion, and the *Saturday Review,* the *Bulletin of the Atomic Scientists,* and the *New York Times* endeavor to interpret the real meaning of science to society today. *Scientific American* does a magnificent job in presenting the new thinking across the broad frontiers of science for those with some knowledge of its language.

I fear we must conclude that by and large the scientists who are trying to communicate with "humanists, jurists, philosophers and theologians" are just not getting through with their full meaning. Communication implies at least a working knowledge of the language, of mathematics, and of the great central principles and the philosophic background on which these are based. Among this major group of citizens who still respond almost instinctively to a philosophy of "final cause," some examination of science and its increasing rationalism would serve to broaden our total social horizons immeasurably.

In closing this discussion, I think we must recognize the acute danger of a degeneration in our civilization, through the failure to comprehend the character and power of the means at hand for improving the condition of the great bulk of the world's people.

Is it too much to ask the humanist, the man of letters, to put the whole of knowledge in proportion, to de-emphasize the encrusted past and look to the future in light of today's proven rationality? Find the good in our times, not from a fixed, unreal baseline, but from the base of dynamic advance of human thought that can be ours. See God, not as static, but as dynamic, growing with our comprehension, eternally challenging our imagination by the intricate and beautiful

15. Philip P. Wiener and Aaron Noland, *The Roots of Scientific Thought,* Basic Books, New York, 1957.

order of the universe, and spurring us on to find the laws that govern it. Create a philosophy that is not fixed but which responds to the relativistic framework of our contemporary society. Give us means to establish new and updated values that represent the major forces incident in society at a given time—values that have the most rational basis at the moment, values that at least point in the proper direction. Establish new standards of beauty and harmony that are consonant with the realities of our changing existence.

Not long ago I rode the pilot's seat of a great passenger jet that was to whisk me across a continent in a few hours. The statistics were impressive for it was a pterosaur of hundreds of tons, fleet as a sound wave, flying seven miles above an Earth that was hidden by cloud. Man was in a void of cloud and sky and his unaided senses could tell him nothing. But the machine knew—as it passed invisible markers, it turned gently to new courses. The voices of the pilots and engineers, as they checked and recorded the instruments at each minute, formed the rhythm for the complex symphony played by the machine. As instruments brought instructions from the ground, the beat of the engines changed, engines so powerful that a few centuries ago they might have powered almost the whole of man's industry.

As this fleet and powerful bird began its descent, the background sound of the engines diminished, orders followed in quick succession, the staccato rhythm of the crew's checks increased in pace and dominated the scene. Gently and safely this great machine nosed its way through the cloud and settled to the earth at the tiny spot where it had been told to go. This was a drama of our time, the theme of a ballet, symbolic of the submission of a great intelligent machine to man's will—to his touch and command. Truly, no king of any age could have found such a responsive slave.

Yet, to command the obedience of this slave, man himself

must be disciplined—he must instruct the machine in a precise and orderly fashion, employing a skillful coordination of muscle and brain. To exercise this power of command, man must submerge his intuitive reactions and beliefs to the cold scientific logic of the situation, for the machine expects its own standards of perfection from its master. If these standards are violated, the machine becomes unruly and, like any great monster, it may fight back and even perhaps destroy its master and itself.

This is the never-ending discipline of mind and body that our scientific culture now imposes. In return, it holds the promise of untold rewards.

# 5. A Strategy of Maturity

Carved in the stone of the Great Hall of the National Academy of Sciences are these timeless words of Aeschylus:

Hearken to the miseries that beset mankind. They were witless erst and I made them have sense and be endowed with reason. Though they had eyes to see they saw in vain; they had ears but heard not; but, like to shapes in dreams, throughout their length of days without promise they wrought all things to confusion—

They had no sign either of winter, or of flowery spring, or of fruitful summer, whereon they could depend, but in everything wrought without judgement until such time as I taught them to discern the rising of the stars and their settings. Aye, and numbers too, chiefest of sciences. I invented for them and the combining of letters, creative mother of the muses arts, wherewith to hold all things in memory.

Twas I, and no one else that contrived the mariners flaxen-winged car to roam the sea.

If ever man fell ill, there was no defense, but for lack of medicine they wasted away, until I showed them how to mix soothing remedies wherewith they now ward off all their disorders.[1]

1. Aeschylus (525–456 B.C.), *Prometheus Bound.*

Truly, the characteristic of civilized man that distinguishes him from all other creatures is his learning, his ability to utilize knowledge to free himself from the vicissitudes of his environment. Knowledge was respected by the ancients as a gift of the gods—it is the ordered synthesis and abstraction of experience.

In contemplating the growth of knowledge and its perpetuation through education, we are reminded that the history of knowledge has been very fleeting. Significant learning awaited the discovery that information could be represented and preserved by symbols, enabling each generation of men to "stand on the shoulders of those before." The beginning of Sumerian cuneiform writing on baked clay is removed from us by only about a hundred successive lifetimes of three score and ten, or about 300 generations. On this time scale, the span of our own lives is decidedly significant—some one per cent of civilized history. In the scale of geologic time, all of civilization is still in its infancy, with who knows what worlds of knowledge yet to be discovered and assimilated.

As civilization advances, the multiplication of knowledge accelerates at an ever quickening pace. Each step forward in learning, with its application, frees man from physical labor and enables him to delve more widely and deeply into the unknown. There is constant feedback in the whole process. Innovation provides new tools that further permit extension of the limits of human thought. As the volume of knowledge grows, the potential new subjects to be explored are enlarged by some exponent, the square or the cube of time. Thus, though merely a hundred lifetimes from the beginning of civilized history, man's environment is now one of ever quickening dynamic change, an environment that no longer remains even reasonably similar over a single lifetime. This places a relatively recent requirement on society to develop a suitable capacity for rapid adaptation to the fast-changing social scene. Only learning on a broad front can provide that capacity.

Prior to the development of unlimited controlled energy in the last century, the advancement of knowledge depended on a small class of citizens who were freed from the drudgery of providing for their own subsistence largely by the labor of a "working class." Through a variety of social devices, ranging from slavery through feudalism and serfdom, the intellectual elite were maintained in various ways; in return they assumed toward society a responsibility for the discovery and perpetuation of knowledge. Whatever we may think about the justice or effectiveness of the social systems involved, we cannot escape the conclusion that in an economy of extreme scarcity, some such device is a necessity. Since the economy of scarcity has persisted right down to our own lifetime—and in fact prevails today in all but the few highly industrialized nations—this custom has by no means disappeared, and in country after country, the responsibilities and burdens of education continue to be borne by a small minority.

Those countries that have encouraged the development of abundant energy and the application of advanced technology to its control have now abolished the traditional economy of scarcity. In such privileged areas even the average citizen finds that in the middle of this twentieth century he enjoys command of the equivalent of a hundred slaves who can produce miracles far beyond the dreams of any caliph. The response has been immediate. Stimulated by the freedom of our American system of natural selection, a whole new class of "learned men" is rising, constituting an appreciable proportion of the population. I remind you in Table 3 of the startling figures from the second chapter.

This complete change in the character of the lettered classes in little more than a generation stems on one hand from the new leisure created by the technological revolution

TABLE 3.

*Explosion of "Lettered Classes"
During the Twentieth Century*

| Year | Baccalaureate Degrees | Doctoral Degrees |
|------|-----------------------|------------------|
| 1900 | 28,000 | 400 |
| 1920 | 50,000 | 700 |
| 1940 | 200,000 | 3,500 |
| 1960 | 450,000 | 10,000 |

—a leisure now largely available to all—and on the other hand from the new obligations for advanced learning which the individual must accept in order to contribute significantly and adapt rapidly to a society functioning in the economy of plenty. This is the extraordinary threshold of the new social revolution in the United States from which we must look to the future. In previous chapters we have examined science and innovation as the sources of our economy of plenty. We have also mentioned some basic sources of potential instability in this new economy:

1. Our educational capabilities particularly with respect to the proportions of learning and skill required to keep it viable.

2. The policies governing the encouragement of science out of which innovation and new technologies are born.

3. The attitudes of society which must provide the underpinning of the new economy derived from a "freer and fuller rationality" acquired from a revitalized liberal education—attitudes that should visualize and measure social stability in a relativistic frame of reference.

4. The potential elimination of poverty by the extension of opportunity for higher education to every level of society,

and by creation of opportunity for that small group of handicapped or mentally retarded who are unable to compete in the selective competition of the new culture.

5. The radically rising imbalance in social opportunity on a planet now unequally divided between the highly industrialized nations and the vast underdeveloped areas—a disparity that is greatly aggravated by the population explosion.

6. The radical change in the character of war and of justice among nations with the introduction of nearly absolute weapons.

In endeavoring to estimate the chances for a stable future, let us turn back to the problems of education. It is perfectly clear that the archaic concept of the university as a place for genteel browsing by an aristocracy has been completely superseded by the social developments of our time. Instead, the university is emerging as Clark Kerr's city of intellect in which the university complex becomes the central and prime purpose of each of the one hundred or so metropolitan areas that are developing as the cores of our national population and strength.

The present effort to achieve intellectual excellence is but one facet of the social revolution. Various efforts are under way to synthesize and interpret knowledge and to take advantage of the powerful new media of communications that our age affords. A certain amount of reaction against new methods by some teachers and school administrators is, however, foreseeable. They see themselves threatened as man struggles to free himself from the shackles of tradition. But the economics of the city of intellect will inevitably impose more efficient and more effective methods of instruction, and especially improved and less demanding forms of administration, leaving the teacher free to concentrate on the more creative aspects of human development.

The idea of continuing education is already on the march. Today the scientist and engineer, and the humanist and theologian as well, find the need for constant intellectual "retreading" in the enlarging horizons of our changing environment. The need goes beyond mere extension or refresher education designed for an agrarian society. The need is for formal, continuing access to courses of learning at and beyond the graduate level, which themselves become minimal prerequisite knowledge at advanced levels.

The human mind, if employed usefully, has enormous capacity, not more than a small fraction of which is ordinarily tapped at present. Moreover, there is more than a suspicion that the mind quickly deteriorates if confined to routine tasks; it simply rusts if allowed to lie fallow. On the other hand, when constantly exercised toward creative ends the mind has indefinite capability for growth (we might note parenthetically that in a million years of natural selection, man's brain has doubled in size). I have heard it said that really great minds acquire the equivalent of a new doctor's degree each five to seven years in their penetration of different fields of knowledge. The widely ranging mind is the obvious counter to overspecialization; it provides tremendous power over the environment without necessarily any loss of depth. Just as our present society glorifies continuing physical fitness, so our future society will emphasize continuing intellectual development by providing educational courses with suitable prerequisites.

The chairman of the board of one of America's great industrial complexes recently told me that to maintain his science-oriented activities at a competitive level, he would be prepared to grant three half-days a week for continued formal education of his scientific and technical staff. He was satisfied that the outworn custom of night classes was past, a custom which, on a continuing basis, taxes man's spirit and encourages superficiality. He wondered how long before each city would be provided with great centers of continuing education from the undergraduate university through the highest levels of learning, employing all the modern methods of communication and teaching now available. This need, already so apparent in science and technology, is likely to spread quickly into the social sciences and perhaps even to the other professions and to the humanities.

Except on a minor and experimental scale, few such formal centers for continuing education are yet in being. The organization of mere discussion groups composed of people of widely different educational backgrounds does not satisfy this growing need. The demand is for depth, successively in each of a variety of fields. One does not need a crystal ball to perceive that continuing education at the graduate level will become a major enterprise of advanced education during the next two decades, thereby accelerating the social revolution so graphically illustrated by Table 3. Continuing education is perhaps one answer to the greater leisure that automation creates. Among other things, it may stimulate widespread and renewed interest in the humanities as a part of the total realm of human experience. For in their adult years "men turn to the art, to disinterested reading, in short, to self-cultivation as a means of keeping their souls alive."[2]

As we turn to professional education, it is not hard to predict that the controversy concerning its form and content will continue. There is much to be said for the view that professional men should, as a prerequisite, be exposed to the broad outlook and flexible thinking inculcated by the liberal arts. But at the same time one must respect the enthusiasm of the student who insists on early access to his field of interest before his creative instincts are dulled. It would be a pity to deny the impetuous student the opportunity to test his mettle by participation in actual experimentation while his intellectual zest is at its height.

2. Jacques Barzun, address to Hofstra University, Dec. 12, 1963.

One suspects that the solution will emerge from a number of current trends:

1. The encouraging improvements at elementary and secondary school levels of curricular content and increased competence in teaching and counselling, which are beginning to take advantage of a whole new array of teaching devices and textual improvements. The result is more rapid progress related to individual capability and consequently to higher standing.

2. Improved articulation between high school and university, with the result that certain elementary university courses are being replaced by greatly improved courses at the secondary level. Some students are now beginning to enter the university with adequate preparatory work. The National Science Foundation reports that in some instances students who have had the benefit of revised course content in the sciences are now being admitted to the university with sophomore standing in these fields. Therefore, the universities must rethink their own curricula.

3. Opportunity for research activity from the very beginning of the student's career to satisfy his creative and manipulative inclinations.

4. More effective organization and presentation of the idea content at the university level, with the finer details for later specialization, and problem solving directed to an ever broadening range of situations.

5. The shift of full professional emphasis and specialization to the graduate school.

As the liberal arts curricula regain their virility, incorporating a reasonable content of mathematics and science organized and presented in effective and palatable form, one can anticipate a rise in the influence of the university-college as a preparation for the graduate schools in both the humanities and the sciences. Indeed, this reintegration at the undergraduate university level can be hastened by recognition of the essential need for creative research outlets at every level of education, and by the awareness that society as a whole now requires the powerful techniques of mathematical symbolism and logic.

Perhaps the most far-reaching revolution in American education in the next decade will be the emergence of the junior college as the major development of our advanced education. This institution is destined to affect more citizens than any other. With the elementary and high schools now able to offer their students advanced standing, the two-year junior college becomes potentially capable of turning out the engineers and advanced laboratory technicians that formerly required the conventional four years of college training. Basically, the junior college should not be conceived as an intermediate stepping-stone to the university or a "bob-tailed" university-college. Rather, it should be thought of as a two-year terminal institution in its own right. Some junior-college students, of course, will find themselves intellectually and go on to higher training, but the junior college is best dedicated to its two-year graduands. In this scientific age, every city of 50 or 100 thousand population will need such an institution. In the metropolis it becomes an essential part of the city of intellect. As much as 25 per cent of our population will probably travel the college or university route, but we must expect nearly half our population to receive advanced training at the junior college of the future. The complexity of our fast-moving technology creates an urgent demand for a sufficient number who can manipulate it.

A profound evolution in the character of American business is emerging from the revolution in education, illustrated by Table 3, as a result of the broadened educational base of its managers and participants. It is important that the worth of this evolution be recognized because it is making a major contribution to the economy of plenty. The old image of the businessman as a robber-baron has all but vanished from the contemporary scene. American business today enjoys a level of respect which should be more realistically recognized by our educational system. Indeed, failure to give credit where credit is due may seriously retard our social and economic progress. For American business, employing radically new methods and procedures, plays an important role in the transition between science and technology and the exploitation of innovation for human benefit.

Successful business has had to adjust to the rapid changes of our time. Products and production methods do not remain static for very long; the traditional methods of the last century have almost wholly disappeared. A new breed of highly educated planners and managers has appeared, who have strong technological orientation and sophisticated and formalized methods of management and control.

The care with which the strategy and tactics of the modern company are planned, the drive toward new, better, and more reliable products at lower cost, improved marketing techniques, analysis of opportunities through creation and demonstration of new customer needs, and a sensitive response to the advancement of the national economy are apparent in the work of the American Management Association and the National Industrial Conference Board. By its willingness and ability to accept change, American business has played a crucial role in the transformation of our national economy. Stimulated by its business leadership, the United States is making the transition at least two decades ahead of most nations—for many of them the major headaches are still in the future or just beginning.

It is of considerable interest to observe that, under the principles of natural selection (which operate in economics

113

as well as in the natural sciences), American business has learned that the greatest success competitively (a) does not come out of the hide of the worker, who is also the customer; (b) does come from highly educated levels of management and technology; (c) can achieve constantly increasing artistry in its products and methods; (d) is able to profit from intimate contact with, and support of, education at all levels.

The extraordinary degree of artistry in merchandising that is so successfully enjoyed in the highly competitive mercantile industry by such leaders as Stanley Marcus foreshadows the increasing articulation of culture into our day-to-day affairs. The better forms of American advertising are both effective and pleasing in their presentation and educational content; they play an essential role in influencing the public audience to more progressive buying attitudes.

In certain areas of advertising such as radio and television, where a severe limitation on services by natural restriction on available wave frequencies leads to a quasi-captive audience, one must withhold final judgment. The infantile and unartistic approach in these otherwise powerful communication media can be established by the most superficial observation. This retrogressive form of advertising, which tends to exploit rather than to educate its audience, is not representative of the capabilities of the art or of the medium. Here we must await the development of some degree of maturity, fortified by the expectation that in the long run a successful enterprise will learn to lead and not to follow. Customers sometimes have better taste than they are given credit for, particularly in a society whose education is rising so quickly.

I emphasize the remarkable evolution of American business because there is insufficient understanding and appreciation of how American business today contributes to the world economy, of how decisions are made by enlightened leadership, and how business statesmanship has emerged out of today's commercial competition.

The rise of the Communist governments has shown that in any system one must have some form of industrial management which ultimately turns out to look surprisingly like the "company." What is needed are thorough comparative studies, stated in comprehensible terms, of the Communist system, the European cartel and its closely related socialist systems, and our own new system as these have evolved in the last three decades and now function in the several economies. Such a study might include a range of subjects like capital formation in suitable proportions, long-range planning in a changing economy, decision making, management, production methods, assimilation of science and technology, marketing processes, and employee relations and benefits.

The principle of diversity, inherent in a system of free enterprise, is emerging as the dominant factor in the evolution toward a science-based economy. The planning function, for example, is far better and more imaginatively performed under the forces of natural selection in the free enterprise system than it is under state socialism. To extrapolate plans very far ahead in these days of extraordinary technological change requires the most intimate association with, and knowledge of, every intricate level of today's industry. This is not a responsibility that can be successfully assigned to leaders selected politically. The failure of the average citizen to comprehend today's business is illustrated by the prevailing attitude toward the function of depreciation in a highly technological society. In the early days of the industrial revolution, depreciation was related to the physical life of a tool or factory. An item was fully depreciated only when it was worn out. The customer came to look on more rapid depreciation as an artificial and unwarranted charge against the product. But today, *technological obso-*

*lescence* determines depreciation. If the telephone company depreciates its exchange too slowly, it simply delays the time when the customer can enjoy intercity dialing. If a given electronic device is not depreciated this year, it postpones for another year introduction of a new device that will be superior either in providing added capability or lower cost.

The judgments necessary to the regulation of today's technological industry are often value judgments of extreme intricacy which can easily be attacked by the demagogue before an uninformed public. The result hampers the economy and deprives the public. Natural selection based on diversity can operate successfully only when the judges are reasonably qualified. That is why public appreciation of technical value judgments is essential.

I have dwelt at some length on the evolution of American business, since its role is so central to the future city of intellect. Business today reacts with our institutions of education at every level. This constant interchange and interaction is becoming a part of our new civilization. Kerr summarizes:

> The campus and society are undergoing a somewhat reluctant and cautious merger already well advanced. MIT is at least as much related to industry and government as Iowa State was to agriculture. Extension work is really becoming "lifelong learning." . . . The student becomes the alumnus and the alumnus becomes the student; the graduate enters the outside world, and the public enters the classroom and the laboratory. Knowledge has the terrifying potential of becoming popular, opening a Pandora's box.

In the face of these new forms of interaction, education is wracked with problems. The *New York Times* (January 2, 1964) editorially decries the frequent absence of the professor from the classroom, with its consequent disservice to the student. But it fails to mention the benefits from that absence, in better government, more desirable social development, or improved products and services. The professor's responsibility is to the whole community, and he should be free to pursue such activities as will maintain his optimum competence.

The very shortage of skilled masters during the next century warns that we should look to sources beyond the traditional university faculties for our professors. Some of the finest scientists and technologists are to be found in the great institutional laboratories of industry, of government, and of the foundations. The skills of these men should be tapped, through part-time appointments at universities, through research supervision of students in their own laboratories, or through other appropriate mechanisms, in order that they may perpetuate their kind through student instruction and counselling. I believe that the really enlightened industries will now generally cooperate in such endeavors. Arrangements must, of course, be carefully circumscribed and formalized to maintain the highest standards of academic achievement. At the top level of graduate communication there is no substitute for the intimate master–student relationship. Therefore, to meet the burgeoning demands for vastly enlarged graduate opportunity, the interests of society now require that all highly educated men contribute, in part at least, to the educational process. The complex interrelationships among industry, government, and education can be of immense benefit to all. But, of course, we shall need to create standards of behavior that will minimize the obvious problems that must ensue.

Quite clearly, in changing our society in the twentieth century from an agricultural orientation operating within a traditional economy of scarcity to a science-based orientation functioning largely within one hundred or so metropolitan areas, we must expect most violent forms of social debate. Each measure or experiment will inevitably encounter the jibes of the uninformed who can see only the superficialities. This reaction probably imposes a useful discipline on those carrying responsibility, but it contributes little of substance to the reconstruction of society. And then there is the political extremism of both the far right and the far left that is the inescapable expression of social resistance to change. Existing social institutions do not readily capitulate in the evolution toward a more mature society; in their opposition they glorify some particular ideal or procedure, dredged up from the past, to the exclusion of all other relevant matters. Typically, extremism endeavors to mold society into some greatly oversimplified theoretical pattern. Almost without exception, extremism endeavors to stifle debate and evaluation of alternatives by the authority of its own glorified ideal. Extremism resembles a curious psychological phenomenon that is commonly encountered in the design of electronic circuits. Such circuits can be readily arranged so that no matter what impulse is applied, only one narrow form of signal emerges. We might say that such a circuit has been "brainwashed," since it is unable to discriminate among alternatives. No matter how sophisticated the range of inputs, the same old line comes out; like a resonance cavity it responds only to one selected signal.

This psychological phenomenon of extremism reached a high point in the early days of Communism, and it has a long history in religion that has by no means completely died out. Late in 1963, for example, massive riots at Srinagar, Kashmir, were triggered by the theft of a hair of Mohammed.[3] In their endeavor to shut off knowledge, the Societies for Prevention of Evil and Encouragement of

3. *New York Times,* Dec. 30, 1963.

Virtue raid and close the Saudi schools for girls as fast as they are opened by Prince Faisal.

The undiscriminating and unvarying Russian response of "American imperialism" and the single-minded Bircher charge that every difficulty derives from a Communist plot are typical of political extremism. This curious psychological phenomenon appears to clamp off the creative channels of the brain that associate ideas in various ways and evaluate alternatives maturely, and a kind of premature senility is thereby induced. Because this type of mental aberration is capable of widespread propagation by the advocates of moribund social institutions, extremism is certain to delay the achievement of a mature society and unnecessarily perpetuate some aspects of the economy of scarcity. As Keynes has observed: "Practical men, who believe themselves exempt from any intellectual influences, are usually the slaves of some defunct economist."[4]

As we approach the problem of poverty, I am reminded of Thomas Hobbes' description of life in the seventeenth century as "poor, nasty, brutish, and short." Today, poverty in our nation is imposed largely upon the inadequately educated whose contribution to the economy is thereby radically reduced.

The alleviation of poverty has some interesting aspects, particularly because the uneducated undoubtedly include a very large group of potentially competent—even brilliant— individuals. Unfortunately, there still is, in many communities, a tradition of positive resistance to education:— "All the boy needs is a good Christian training;" "Education will start them on the path to sin;" "Let them use their hands like me." Such attitudes were tragic in an agrarian economy; today they are dangerously antisocial. In de-

4. John Maynard Keynes, *The General Theory,* Harcourt, Brace, New York, 1936.

pressed areas the schools are characteristically minimal and are often quite deliberately kept so by the community. This attitude accounts, in part, for the so-called "pockets of poverty." Moreover, such attitudes seem to intensify in proportion to the distance from a good graduate center, thus suggesting the strong influence of first-class higher education on community attitudes and levels of achievement.

I would doubt that the sentimental approach of the do-gooders can have much effect in the long run on the elimination of poverty. Instead, some really good quantitative operational research into the problem could identify the basic underlying factors as well as the first-order social and economic forces that must be reckoned with if poverty is to be overcome. The elimination of poverty calls for a new moral attitude—the obligation to pursue one's education, training, and skills to the full limits of one's capabilities. To fail to do so reduces the contribution that the educated entrepreneur can and should make to society. Moreover, the person holding a job below the level of his potential competence blocks that job for one less competent. The virtue of thrift takes on a new meaning—the husbanding of ideas and skills and their judicious application.

Thus the fight against poverty relies basically on adequate education at every level and counterattack against the forces of reaction that block the realization of this goal. Already the drive toward excellence in the elementary and secondary schools has begun. The unwise concept of reducing all education to the level of the mediocre in the name of democracy is finally giving way to the original American ideal of freedom of opportunity.

Of course, there will always remain a small class of retarded or handicapped unfortunates for whom education holds little hope. It is possible, however, to find promise for their welfare. The dole is not the answer, for such people often take pride in even limited accomplishment. Here the efforts of such groups as the President's Committee on Opportunity for the Handicapped, organized by Major General Maas (now deceased), deserve every support.

Intimately related to the problem of poverty, and far more dangerous than the atomic bomb, is the frightening spectacle of the population explosion. People simply do not understand or are unwilling to face the stark facts of population arithmetic. Motivated by humanitarian considerations, man has applied science to the eradication of disease, but he has at the same time thrown his ecology terribly out of balance. The fact is that in the next generation we will add more population than the world already has. Moreover, even if the problem is successfully attacked, the world's population will continue to rise for another century. Consequently, by the year 2200—which is perhaps the earliest that effective controls could be achieved—the population will potentially rise to more than 20 billion, some seven times the present density.[5] I say potentially because such population pressures predispose the world to epidemic disease or to vicious power conflicts that could release an unthinkable nuclear war. Quite apart from the ultimate dangers, the present population rise in many parts of the world is outrunning man's capacity to provide education, capitalization, or even the minimal human amenities.

The United States National Academy of Sciences has courageously reported[6] clearly and succinctly on this critical danger. Common sense, strongly buttressed by scientific analyses, calls for world action to revise the moral concepts derived from the vastly different circumstances of the past

5. See Reports of the Population Reference Bureau, R. L. Cooke, Director, Washington, D.C.

6. *The Growth of World Population,* Publication 1091, National Academy of Sciences–National Research Council, Washington, D.C., 1963.

and to bring under control this most critical of all man's problems. We must applaud those few, like John Rock, whose voices are raised loudly on this vital issue:

> Our nation still has time to throw its rich resources and skills into a constructive program to help mankind bring its overabundant fertility under rational control. How much longer that choice will remain open to us, however, is uncertain, and the more rapidly we decide to exercise our option, the better for all of us.[7]

Perhaps the most compelling challenge facing us today is the provision of opportunity for the underdeveloped peoples of the world. No problem is so inextricably linked to the worst elements of the past—indeed it is the voice of the past re-echoing in our own day. Moreover, though it cannot be proven, one suspects intuitively that there is a limit to the imbalance of opportunity that the world can tolerate as the economies of the industrialized nations continue to rise in the face of the deepening misery in other parts of the world. How and at what time or at what level this limit will be imposed, one cannot foresee. Perhaps the limit is economic, arising from a growing price differential in basic commodities or from the mere cost of our more sophisticated products. Perhaps the limit is political, and envy and hatred directed at the privileged status of our society will result in reprisals that will drag our economy down. Perhaps the limit is military, with an ever growing wave of revolt and unrest sweeping uncontrollably over the world to engulf us. Perhaps the limit will be imposed by some unpredictable combination of all of these.

The dangers of this growing imbalance led the Committee for Economic Development to point out that: "The greatest

7. John Rock, *The Time Has Come: A Catholic Doctor's Proposals to End the Battle over Birth Control,* with a Foreword by Christian Herter, Knopf, New York, 1963, p. 203.

122

economic problem facing the United States is not how she can raise her own standard of living but how she can harmonize her economic development with the world-wide process of growth."[8] Quite curiously, as a result of our economic achievement in creating from science an economy of plenty, we have become a nation of privileged status among nations, in much the same position as were the privileged individuals of old. We have inherited as a whole nation all the problems of those often hated individuals of the past and, at the same time, all the same responsibilities. The entire problem is aggravated by the population explosion, arising from the export of our newfound health but without concomitant conveyance of the means of serving the "poor" who are thus multiplied.

The dangers as well as the essential immorality arising out of the disparity between the have and have-not nations have been recognized by our political leaders in connection with the explosion of our own economy since World War II. It is said that we have expended foreign aid of more than 100 billions of dollars since that time—more than a whole annual federal budget. Our foreign aid has generally taken the form of resources put into the hands of the nations concerned, to be allocated largely under their own management. This is done on the hypothesis that each nation knows best its own problems and how best to manage its own affairs. The only important exceptions were President Truman's Point IV program, under which part of the funds were devoted to technical assistance, and the technical military assistance accompanying some of our military aid.

The results have been varied, and their pattern forms an interesting study. The Marshall Plan was a historical achievement of the first magnitude. Our aid went to nations with high scientific and technological skills, and with it they

8. *Problems of United States Economic Development,* Vol. 1, Committee for Economic Development, New York, 1958, p. 69.

have almost duplicated our own economy of plenty in a virile postwar Europe. In a mere two decades they have arisen from the war's incredible destruction to join us as privileged nations. Our military aid to NATO was equally successful. Managing an alliance of fourteen nations is a neat trick, but it has been done with extraordinary understanding and success. NATO has bridged a very dangerous time gap in which the possibility of another world conflagration was very real. And, despite inevitable problems, NATO still functions well as a major tool of Western policy.

The major proportion of our aid to the underdeveloped world has enjoyed no such success, and we perhaps should not have expected dramatic results. Yet the self-interest of all suggests that such aid is a failure if it does not at least narrow the widening gap of disparity. But the disparity is seriously and visibly growing, year by year.[9] In contrast to this failure is our growing and almost unique success in aiding rehabilitation of Puerto Rico.

Experience tells us that we just have not yet learned how to give effective aid to the have-nots. It is easy enough to blame the failure on reasons which may or may not be justified. Some say the funds are dissipated by irresponsible officials of weak nations that cannot control their leaders. Others say that without education on a sufficiently broad base, money spent to trigger a modern economy is useless. Still others suggest that we are trying futilely to supply basic needs directly, and to do this would take several times our whole economy. Some blame the Communists. Others level the finger at our own inept administration of the program and our failure to pursue Point IV. Regardless of the merit of the various criticisms, *the first-order fact is that none of*

9. See Eugene R. Black, 1963 commencement address, MIT (*The Technology Review,* July 1963), who comments: "the gap between the thousand and hundred dollar countries does not seem to be closing. In fact, it is probably growing wider."

*us has yet learned how to do the job.* And we had better find out quickly. I emphasize the importance of a major attack, because it is too late to permit further experimentation by nonscientific theoreticians whose ideas have no more established validity than those of the man on the street. And it has also become too expensive.

The military have faced strategic problems of this magnitude many times since the war and, with their new science orientation, have developed a number of devices to deal with them. One is to bring together great leaders of thought for many months to analyze each aspect of our experience (country by country if you like), and then weave the pattern together analytically. (The learning time for man is about the length of one university semester; he must examine a field of knowledge that long to comprehend it.) Another approach is the initiation of high-level operational research. Still other forms of study and analysis have been well tested.

The point is that as a nation we have developed the techniques to solve tough problems. The time has come to turn our attention to the problem of the underdeveloped areas in a big way—and in so doing to use every iota of experience that we have acquired in problem solving. All the ideas, theories, and suggestions, as well as tested experience, should be subjected to the intellectual rigor of analysis and study, so that a more viable plan of administering and effectuating our foreign aid can be evolved.

What might come out cannot be predicted, though it might be surprising. One suspects that the solution to this dreadful problem may involve a compassionate, people-to-people approach at every level—education, technical aid, leadership in business management, and organization. Simply to send money, without the master–student relationship, is both too cold and too uncommunicative of the skills required. To try to superimpose the pattern of our own development in this science-oriented age would leave strug-

125

gling young nations far behind and noncompetitive. Yet somehow our program of aid should draw on America's greatest strengths—its educational potential, its technological know-how, and its tremendous acumen in business which now draws these things together to create wealth for any people at an optimum rate. The Peace Corps is only the merest beginning of a genuine people-to-people approach to this difficult problem.

With the greater leisure available to us, our own leaders of thought and method must perhaps dedicate some portion of their lives and effort to the solution of this problem and to the development of these peoples. In particular, with its analytical methods and its scientific philosophy, science might well turn a greater proportion of its efforts to this problem—though to do so effectively would require a substantial measure of government recognition of the part that scientists might play both in the analysis of the problem and with respect to its substance. Men who are close to the problem, like Black, have suggested trying such a course:

> One of the most brilliant successes in international cooperation since the war was one organized by scientists, —the International Geophysical Year. . . . It seems clear that the scientific community is the best possible source of inspiration and of planning for the rational and energetic attack on the technical gap separating the developed and the underdeveloped nations. [10]

Finally, I would comment that in our struggle for a mature society, man now has in his hands the tools to prevent war. For the first time in his history, the power of the nuclear weapon—managed by a mature society—stands between the ordinary citizen and the chauvinist, buccaneer, and adventurer. The potential aggressor may make loud noises, but these cannot jam the clear and penetrating signal of the

10. Ibid.

atomic stockpile. As Reston notes in his New Year's Day assessment:

> There is a significant downbeat in the tone of the New Year statements out of the major world capitals. They are moderate and sensible for a change, and almost wholly devoid of the strident threats, boasts and grandiose schemes that have greeted most of the post war years . . . This new tone did not happen just because the world's leaders got tired of the sound of their own voices. A single momentous fact has now been widely accepted. This is that no country can now launch a major atomic war on another atomic power without assuring its own destruction. The idea of sudden victory by a devastating first atomic strike has lost its appeal.[11]

The management of the ultimate weapon has brought heavy responsibility but with it a deep sense of military maturity. The infantile parade of military power at each real or imagined insult has given way to serious review of the legitimate interests of both contestants.

This is not to say that immense world problems do not exist or will not arise. A great multiplicity of international political problems in every part of the world today stems from the nuclear power that now prevents the employment of a major war by any nation as an instrument of its policy. We have yet to learn the rules whereby such problems can be settled most readily now that man no longer has resort to major conflict. But we are approaching their settlement with the recognition that resort to the total wars of the past is no longer a feasible alternative. Today, responsible men are constrained to let time cool the fever of excitement and lend perspective to their thinking.

We are awakening to the realization that the foreign policy of any nation has two quite separate and sometimes mutually

11. *New York Times,* Jan. 1, 1964.

incompatible components—the achievement of immediate policy objectives in terms of what seems most desirable and necessary at the moment and the creation of an atmosphere of international confidence, trust, and respect in which problems can be more readily solved with the passage of time. This second facet now looms with major emphasis as the nuclear warhead removes the potential of total war as the central instrument of national policy. Yet this fabric of trust can be woven only out of a multiplicity of threads of mutual experiences in the realm of business, finance, education, science, religion, art, and literature. Intergovernmental relations can but clear the way and encourage the weaving of that fabric by the people themselves. This takes a long time, great patience, and wisdom.

In this atmosphere military forces take their proper place as an important means to maintain, police, and enforce the peace but no longer as the sole, ultimate tool of national policy. At the moment, in absence of agreed means of international jurisprudence, our international justice unavoidably assumes a vigilant flavor of immediate reaction. This is inevitable if peace is to be maintained under circumstances that would avoid the ultimate frustrations that breed total war. Yet, even as procedures of international justice emerge, effective military power in being will be required for enforcement, for no system of justice is meaningful that does not implement its effectuation and protection.

Our consideration of the many complex interactions of science and technology with every part of the social structure leads us to conclude that in order to maintain, enjoy, and enlarge our extraordinary economy of plenty, we must adopt a strategy of maturity. The continued growth of our society in this scientific age demands the conscious and thoughtful effort of all of us. In this age of science, the key to a stable society lies in its ability to discern and to direct its prime attention to the real issues of the time—issues that in the long run will have first-order impact on its genuine well-being. Diversionary activities, encouraged by demagogues and extremists and nourished on prejudice, can flourish only in a society incompetent to form intelligent judgment on basic issues and ignorant of the elements that encourage growth.

A strategy of maturity is, I believe, a social strategy arising out of a rational attitude of mind that evaluates alternatives objectively and tests each step analytically. It is a strategy designed to capture and explore the optimum advantages that our scientific culture affords. This is a strategy based on the power of a nation comprising a hundred cities of intellect, whose business, educational, and cultural activities interact strongly at every level through an educated citizenry in their manipulation of the most sophisticated technological skills. It is a strategy which recognizes each social problem in accordance with a scale of its reality and seeks solutions based on moral, ethical, and political principles developed in consonance with the environment. This is a maturity which recognizes that science and engineering are the manifestations of our time that have given our age its character.

Man now sees—almost within his grasp—the promise of a society beyond the dreams of the sage and poet. Can he rise to accept the challenge of maturity?

# Index

134